P9-BYH-591

DATE DUE

		PRINTED IN U.S.A.

Is Shame Necessary?

Is Shame Necessary?

New Uses for an Old Tool

JENNIFER JACQUET

Illustrations by Brendan O'Neill Kohl

PANTHEON BOOKS

New York

All rights reserved. Published in the United States by Pantheon
Books, a division of Random House LLC, New York, and in
Canada by Random House of Canada Limited, Toronto,
Penguin Random House companies.

Pantheon Books and colophon are registered trademarks
of Random House LLC.

Library of Congress Cataloging-in-Publication Data
  Jacquet, Jennifer.
    Is shame necessary? : new uses for an old tool /
  Jennifer Jacquet.
        pages   cm
    Includes bibliographical references and index.
  ISBN 978-0-307-90757-8 (hardcover : alk. paper).
  ISBN 978-0-307-90758-5 (eBook).
    1. Shame.   2. Guilt.   3. Human behavior.   I. Title.
  BF575.S45J335   2014   302.3'5—dc23   2014020331

www.pantheonbooks.com

Illustrations by Brendan O'Neill Kohl
Jacket image: Vetta/Getty Images
Jacket design by Janet Hanson

Printed in the United States of America
First Edition

9   8   7   6   5   4   3   2   1

Shame. The feeling that will save mankind.

—Screenplay for Andrei Tarkovsky's *Solaris* (1972)

Shame is for sissies.

—BARON EDWARD VON KLOBERG III, American lobbyist
  (1942–2005)

Contents

Is Shame Necessary?

1

Shame Explained

> [A man's] moral conscience is the curse he had to accept from the gods in order to gain from them the right to dream.
>
> —WILLIAM FAULKNER, interview in *The Paris Review* (1958)

In 1987, thirty-year-old Sam LaBudde walked into the offices of the Earth Island Institute, in San Francisco, looking for a job fighting rainforest destruction. He walked out headed for Mexico, planning to become a spy. In Earth Island's lobby, LaBudde read an article about how the tuna industry was killing millions of dolphins in purse seines—large nets that encircle a school of tuna and are then drawn tight to catch everything in their "purse," including dolphins, which then drown or are crushed in the gear that pulls in the net. The article was powerful, but there were no visuals. Instead of saving the rainforest, LaBudde convinced Earth Island to send him a video camera (this was the 1980s, before camcorders were common) and he set out to find a job on a tuna boat, as a ruse to collect footage of the dolphin slaughter.

LaBudde succeeded in becoming a deckhand and later a cook on board a Panamanian fishing boat operating out of Ensenada, Mexico. At great personal danger, he filmed several tapes of dead and dying dolphins caught in the tuna-fishing gear. Earth Island used the footage to launch a media campaign on national and local U.S. television. It was written up in newspapers and magazines, including a three-part series by Kenneth Brower in *The Atlantic Monthly*. The campaign was based on shaming, which involved exposing the transgressors to the American public. The target of the shaming was the tuna industry, specifically the three largest tuna companies: StarKist, Bumble Bee, and Chicken of the Sea.

Around this time, I convinced my mom to buy me a book titled *50 Simple Things Kids Can Do to Save the Earth* (1990), and on the book's advice I wrote a letter of inquiry to the Earth Island Institute. Weeks later, I found a response from the same group LaBudde had visited in San Francisco in my mailbox on our cul-de-sac in Ohio. The envelope included a haunting black-and-white photograph LaBudde had taken of a dolphin hoisted and dead on a tuna vessel. The campaign materials Earth Island sent me had shamed the tuna industry, but for me they incited guilt. Regardless of what anyone else thought, my conscience told me that what was happening to dolphins was wrong (never mind what was happening to tuna). Guilt is a feeling whose audience and instigator is oneself, and its discomfort leads to self-regulation. Looking at that image was my first time feeling miserable for a creature I had never met—seen only in the pages of wildlife magazines—and my first time, but not last, feeling guilty for something I had eaten.

I needed to do something. At nine years old, I had already learned what the 1980s taught as the new rite of passage: to alleviate my guilt as a consumer. I insisted that our family stop buying canned tuna, and I wasn't alone. The evidence of mangled dolphins saddened and outraged people around the world and incited a large-scale tuna boycott by households and, later, a change of protocol by the major tuna companies. In an interview at that time, Anthony O'Reilly, former CEO of Heinz (which owned StarKist), said, "I think it would be a poor CEO that was not attentive to his customers. And because of the affection that children have for Flipper and because of the gross scenes that were shown in the LaBudde film, there was a growing barrage of criticism—well orchestrated—which I think served to convey a growing sentiment among schoolchildren that the previous fishing methods were no longer acceptable."

Those schoolchildren and their parents were among the relieved shoppers when the "dolphin safe" eco-label was introduced. We all felt better and resumed eating tuna. I didn't think about the tuna problem or the dolphin-safe label again for more than a decade. When I finally did, it was because I realized we had been duped.

The dolphin-safe logo, introduced in 1990, was but one of the newly introduced market tools to save the world. That same year, the U.S. government established organic food legislation (although the first organic food certification was founded in Santa Cruz, California, in 1973). The Forest Stewardship Council, an international scheme, was formed in 1993, after years of discussions about how to certify forestry practices as sustainable. The Marine Stewardship Council, the main program to certify sustainable fisheries, started up in 1997, the same year that Fairtrade International was created. Many more labels followed.

Until this push for certification, the goal of shaming campaigns and boycotts had been to fundamentally change entire companies or industries. Activists like Cesar Chavez, behind the strike and boycott of table grapes in the 1960s, would not have ended their efforts with a label on grapes that read, PICKED BY FARMWORKERS WHO EARNED A MINIMUM WAGE. The aim was not to satisfy the concerns of a few consumers, but to (among other things) change federal rules for the minimum wage and workplace safety for all farmworkers. The exposure of poor sanitation in meatpacking factories in the early twentieth century was not intended to produce a label that would allow concerned consumers to purchase hygienic meat— the goal was to raise sanitary standards everywhere.

But by the 1980s, the notion of directly changing supply was being displaced by the idea of changing demand. At first glance, this strategy seems reasonable, particularly in a laissez-faire economy: if demand changes, supply should respond. The two are balanced on opposite sides of the ledger, as I was often reminded during my six years studying economics. The new plan of engagement, which even environmentalists supported because the Reagan era had ushered in a political climate that had become more and more hostile toward regulation, saw the pocketbook as the most powerful avenue of persuasion. This strategy allowed consumers to retain their "freedom to choose" (the battle cry of libertarian economist and free-market popularizer Milton Friedman) and suggested that guilt-

ridden shoppers could avoid their uneasy feelings by simply changing their buying habits.

As the focus shifted from supply to demand, shame
on the part of corporations began to be overshadowed
by guilt on the part of consumers—as the vehicle for
solving social and environmental problems. Certification
became more and more popular, and its rise quietly suggested that responsibility should fall more to the individual consumer rather than to political society. The notion
of certification also disposed of shame and targeted guilt
as the main form of engagement. Guilt could be used to
motivate individuals in a way that it could not be used
for entire industries or suppliers, but it could also be
used *only* on individuals, since groups, such as the tuna-
fishing industry, lack a conscience and therefore cannot
feel guilty. The goal became not to reform entire industries but to alleviate the consciences of a certain sector of
consumers.

But the problems of pesticide use and worker exploi-
tation and bottom trawling cannot be solved with an
individual choice. If pesticides are absent from my food,
but they are in everybody else's, they still leach into our
shared water supply. If I eat dolphin-safe tuna, but everybody else continues to eat dolphin-unsafe tuna, dolphins
remain in trouble. If I stop flying, but nobody else does,
carbon dioxide emissions continue to steeply increase.

The trouble with collective-action problems is that
they are difficult to solve by changing the psychology
and therefore the behavior of individuals. These types of
problems often require larger, often structural changes. It
would not have been sufficient for individuals who understood and felt guilty about ozone-depleting substances to

stop buying them, because these people would have been the minority. To solve the problem of the ozone hole, most if not all of the production of those substances had to stop.

With so many recent collective-action problems, especially those related to labor and the environment, we have been asked to engage with our guilt about these problems as consumers rather than as citizens or activists—not even as organized groups of consumers, which have been responsible for large-scale boycotts, but as individual, household purchasers who make decisions only as individuals. Guilt's power is limited, but it can also be profitable if a niche set of goods and services can capitalize on relieving the ill feeling. Yet when issues reach a moral imperative, it is not sufficient to deal with them at the level of individual choice. It was not enough for people who disagreed with slavery not to own slaves themselves— they saw the need to stop everyone from owning slaves everywhere.

My innocence and the innocence of so many others resulted in our inability to effectively engage with guilt. We schoolchildren didn't just want to save our consciences; we wanted to save dolphins. The dolphin-safe logo alone could not assure the survival of dolphins (the logo very likely helped, but largely as an impetus for regulation, not as a purely market mechanism and certainly not as some panacea of its own), and we should not have been satisfied that the portion of the industry that fed us had changed. Instead we should have found it unacceptable that the bulk of the industry had not. If we had not been pacified by the logos and certifications and enlightened supermarkets, we might have remained upset. For dying dolphins and so many other problems, we might have continued to engage with producers not just as consumers but in the ways that Sam LaBudde had: by using shame.

This book explores the origins and future of shame. It aims to examine how shaming—exposing a transgressor to public disapproval—a tool many of us find discomforting, might be retrofitted to serve us in new ways. We will explore the social nature of shame and of guilt, where these strategies sit in the broader panorama of punishment, and what it means for them to work. We will examine why guilt has been asked to perform a function that it is not quite up to—namely, solving large-scale cooperation problems such as overfishing and climate change. We will find that shame is inextricably linked to norms, and that norms are often changing. We will look for examples of shame outperforming guilt and circumstances under which we might see the value of shame and the purpose it serves. We will explore what can make shame more effective and valuable in a world where we are more interconnected and distracted than ever.

What Do We Mean by Shame?

Exposure is the essence of shaming, and a feeling of exposure is also one of shame's (the emotion) most distinct ingredients and intimately links shame to reputation.[1] For our purposes, an audience is a prerequisite for shame, even if that audience is imagined. While there are personal forms of shame that are experienced privately, this book is about not the shame and inner turmoil you would feel if your father brought home an inflatable sofa (trust me), but the shame you would feel if your friend saw it. This book focuses on the shame that is possible because an audience is exposed to a transgression. Moreover, it is most interested in the public act of shaming rather than the emotion of shame.

Shame can lead to increased stress and withdrawal from society. Shame can hurt so badly that it is physically hard on the heart.[2] But shame can also improve behavior. A 2009 study of 915 U.S. adults found that half could recall at least one meeting with a doctor that left them feeling ashamed, most often for smoking or being overweight. Of those who reported feeling ashamed, nearly half then either avoided or lied to their physician in subsequent meetings to evade any further shame, while the other half said they were grateful to the doctor, and about one-third of the patients said they even initiated improvements in their behavior.[3]

Some people do not feel shame even over the ghastliest of crimes. (In 2011, Reginald Brooks, who twenty-nine years earlier had murdered his three sons while they slept, extended the middle fingers of both hands while strapped to the gurney in an Ohio execution chamber, as his ex-wife and the mother of his children watched through a glass window.) At the other extreme, the sting of shame for some people, even for minor offenses, can be crippling. (Writer Jonathan Franzen blamed shame over his first marriage, sexual inexperience, and general innocence for his decade-long writer's block, when semi-autobiographical sentences made him "want to take a shower.")[4] At its most efficient, a sense of shame can regulate personal behavior and reduce the risk of more extreme types of punishment: conform to the expected behavior or suffer the consequences. The threat of shaming often provokes a fear of feeling shame.

Shame Versus Guilt

In contrast to shame, which aims to hold individuals to the group standard, guilt's role is to hold individuals to their own standards. For cultures that champion the individual, guilt is preferable to shame, because shame means worrying about the group. Guilt is advertised as a cornerstone of the conscience. It needs only an internal voice nagging its owner, sending reminders about how awful violence, stealing, or dishonesty can make us feel.

The anthropologists Ruth Benedict and Margaret Mead were the first to draw a distinction between guilt and shame cultures—they claimed that most Western countries fell into the guilt category while Eastern countries relied more on shame. Benedict's 1946 book *The Chrysanthemum and the Sword,* in which she examined Japanese culture without actually going to Japan (the war prevented it), attempted to show that the Japanese used shame as the primary means of social control. China was later filed in the shame category, too, for, among other reasons, the cultural importance of "saving face."

In the West, however, we tell ourselves with a certain amount of smugness that we have been unshackled from shame's constraints. There are a few reasons that this might be at least partially true, and one has to do with the sense of self. Western cultures are more individualized, leading people to see themselves as independent and autonomous, acting according to one's internal compass, whereas people from Eastern cultures are more likely to describe themselves in relation to others. Western cultures also generally lack the tight-knit hierarchy that probably existed in our prehistoric past and still arguably exists to a greater degree in some Eastern cultures, as anthropologists such as Benedict and Dan Fessler have pointed out. (Yet it's also not surprising that shame in the West is frequently associated with poverty—one proxy for low rank in the social hierarchy.) Also, Western societies tend to have a worldview that encourages tolerance of a greater range of certain behaviors, which means we perhaps more often disagree over which behaviors warrant shaming. Many Western countries have also gotten rid of shaming punishments against individuals, especially shaming by the state. It's probably safe to say that we all prefer to live without the fear of dunce caps, whipping poles, or hot-iron branding. It is even tempting to think of shaming as we might wisdom teeth or Puritan doctrine—as a vestigial sign of something that humans needed in tougher times.

However, as novelist Salman Rushdie reminded us, "Shame, dear reader, is not the exclusive property of the East." When Pope Benedict XVI visited the United States after a series of sexual-abuse scandals in the Catholic Church, he confessed that he was "deeply ashamed." After pulling off Wall Street's biggest swindle and before receiving his 150-year prison sentence, Bernie Madoff told the court that he was "deeply sorry and ashamed." After rapper Kanye West stole the microphone at the MTV Video Music Awards from Taylor Swift, the winner of the best female video category, and declared that Beyoncé (whose hit "Single Ladies" came out that year) had made one of the best videos of all time, West said he was "ashamed." Did these Western icons feel guilty? Hard to say. Did they feel ashamed? Without having measured their stress levels, it's difficult to be sure. What we can say is that, at the very least, they wanted us to think they did.

Where Shame and Guilt Fit into Punishment

Shaming, which is separate from feeling ashamed, is a form of punishment, and like all punishment, it is used to enforce norms. Human punishment involves depriving a transgressor of life, liberty, bodily safety, resources, or reputation (or some combination), and reputation is the asset that shaming attacks. These deprivations can be active, in the sense that something is taken away—such as through capital punishment, prison, torture, fines, and pickets—while other deprivations are passive, such as when something is denied, which is the case with ostracism or the silent treatment. (A survey of two thousand Americans showed that two-thirds admitted to using the silent treatment on someone close to them, while three-quarters said they had been the victim of the silent treatment.)

Humans have devised intricate nonviolent punishments. Charles Darwin, for instance, wrote about tribes in South America for whom long hair was "so much valued as a beauty, that cutting it off was the severest punishment." There is solitary confinement, which in American prisons can last for decades. When my brother and I would fight, my mom used a nonviolent punishment of making us sit on the stairs and hug each other for twenty minutes. Shaming punishments can be violent or, most often today, nonviolent. Again, the definition of *shaming* we are using involves the exposure—threatened or actualized—of a transgressor in front of a crowd. These punishments might be nonviolent, but that does not mean they aren't painful.

Punishment can be inflicted by the person or group

MAP of HUMAN PUNISHMENT

TRANSGRESSION

IMPETUS FOR

IS IT...

ACTIVE

PASSIVE

EXPENSIVE

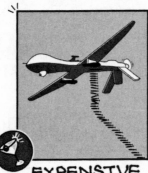

CHEAP

WILL IT BE...

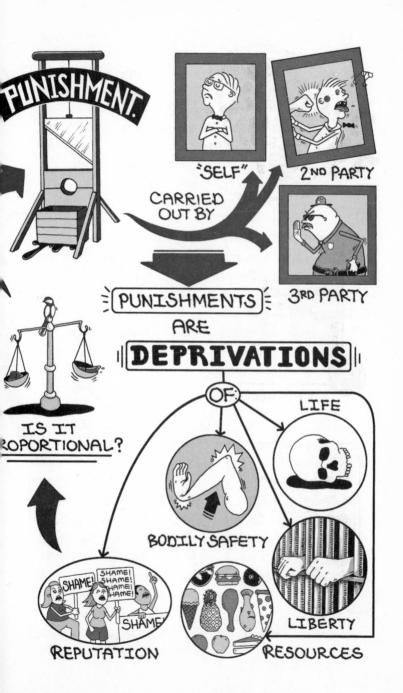

against whom the transgressor transgressed, or by a third party, or by oneself (guilt acts as a form of self-punishment). Generally, punishment carries a cost to the punisher, like the energy needed to perform the punishment, as well as some risk of retaliation. Punishments that are extra dangerous or risky are considered costlier. Sometime in our distant past, we realized that mere exposure to public opprobrium could be used where physical, often violent elimination from the group had previously been required. The emergence of shaming as a social option would have reduced the cost of punishment, because mere exposure that served to damage an individual's reputation in front of the group could have negative consequences—for instance, members of the group might choose not to cooperate with the shamed individual in the future. Shaming and ostracism are closely linked, but shaming is less costly. And unlike transparency, which exposes everyone, shaming exposes only a section of the population.

When and how did shaming emerge? The first hominids, like many other social species, could keep track of cooperation and defection only by firsthand observation. As group size got bigger, and ancient humans grappled with issues of cooperation, the human brain became better able to keep track of all the rules and all the people. The need to accommodate the increasing number of social connections and monitor one another could be, according to the social-grooming hypothesis put forward by anthropologist Robin Dunbar, why we learned to speak. With language, we no longer needed to see someone's behavior to learn about it. Language allowed humans to manipulate social status using gossip, which provided further fuel for a system of reputation and shaming. (And this might not be unique to humans—some scientists suspect that parrotlets, for instance, can identify one another as individuals by their calls and can attach a note of approval or disapproval.)[5] It also meant the crowd concerned with the miscreant's behavior got bigger, because the behavior no longer had to be seen, but could be heard via gossip. Individuals could be exposed to the crowd for transgressing without being physically present.

Negative gossip—a subcategory of shaming—can be considered one of the first lines of defense against a transgressor and was probably as important in human prehistory as it is today. Anthropologists have shown that two-thirds of human conversation is gossip about other people—Polly Wiessner found this to be true in her studies of the !Kung bushmen in Botswana,[6] and Robin Dunbar and his colleagues also found the two-thirds rule held for conversations in a British university cafeteria.[7] Wiessner classified only 10 percent of the conversations she heard as praise; the other 90 percent was criticism, a lot of it in the form of jokes, mockery, and pantomime. The transgressor or one of his close relatives (it was almost always a *he*) was often within earshot, indicating that the gossips expected the verbal shaming would bring him into line. Negative gossip is often employed with the assumption that it will make its way back to the transgressor either directly or indirectly, by influencing others not to be cooperative toward the transgressor.

Spoken language was just the first tool to facilitate gossip. The next communication upheaval occurred with the rise of writing. Since the arrival of writing, there have been, according to Internet scholar Clay Shirky, five major advances in communication technology: movable type and presses, the telegraph and telephone, recorded media, broadcast media, and digital technologies, including the Internet. Each time communication was transformed, shaming was as well. At first we had only gossip among humans that occupied the same physical space; now gossip gets worldwide exposure and can travel via print and digital media, over telephones, television, and cyberspace.

Digital technologies have at once lowered the cost of gossip and exposure and expanded gossip's scope and speed. (No need to call each of your friends; just send one tweet and reach thousands.) Some people even believe the combination of digital technologies will transform culture as much as language did. This could make shame more salient to public life than ever before, especially since the power to use it has been wrested away from opinion leaders and the state and put increasingly in the hands of citizens, which means we should all become cognizant of shame's power and its liabilities.

Take the outpouring of criticism over social media after Susan G. Komen for the Cure, best known for its pink ribbon and its quest to cure breast cancer, announced in 2012 that it was eliminating a $650,000 provision to Planned Parenthood for breast cancer screenings and education. (Planned Parenthood is a provider of legal abortions, and some interpreted Komen's withdrawal of funding as a political statement.) Over the next three days, the Pew Project for Excellence in Journalism tracked Twitter traffic and found 253,465 messages related to the decision to end funding: 17 percent positive, 19 percent neutral, and 64 percent critical. Three days after the Komen foundation's decision, the *New York Times* and the *Washington Post* ran related stories, and the volume of messages on Twitter peaked at 215,383 in a single day (five tweets every two seconds); again, most of the messages were negative. By the end of the day the foundation had reversed its decision. Over the next two days, 64 percent of the messages related to Komen on Twitter continued to be negative, but the quantity of messages fell by 85 percent. Even online, most gossip is critical and serves as a mild form of shaming that attempts to keep people acting in a way that suits the group (whoever the group may be).

Online exposure led the Komen foundation to continue funding Planned Parenthood, and there are plenty of other examples of gossip's power. According to a 2012 *New York Times* article, "Spring Break Gets Tamer as World Watches Online," college students now report worrying about bad behavior being caught on film and immortalized in cyberspace. According to several colleagues, the rise of social media has also meant a decline in hookups during field courses abroad (because students don't want their significant other back home to see an incriminating photo online and realize they're not so significant). But online forms of shaming are also leading to concerns about trivial uses of shaming, disproportionality, opprobrium for the offender (rather than for the offense), and attacks on human dignity.

How Shame Works

In 2010, three colleagues—mathematical biologists Christoph Hauert and Arne Traulsen and evolutionary biologist Manfred Milinski—and I ran experiments testing whether the threat of shame or the promise of honor would encourage cooperation. Groups of six students from the University of British Columbia played a public goods experiment, a setup designed to capture the tension between group interest and self-interest. Each student received $12 at the start of the game and could donate $1 or not to the public pool over the course of twelve rounds. The public pool was then doubled and redistributed evenly to all six players, even those who hadn't donated. This experimental design poses a dilemma familiar to students who are asked to work on group projects in which there is an incentive to free-ride off the generosity of others, yet, if nobody works, everybody gets a bad grade. Everyone in our experiment would gain from public donations, but no one was obligated to give.

In these kinds of experiments, donations generally start out high and then decline over time. Most often, each player walks away with less money than if everyone had cooperated. (In our experiment, every player could have gained $24, but of the sixty students who participated, only one managed to go home with more than $24, and the other fifty-nine students each left with less.) These types of experiments have their limitations—students are given money at the start of the game (which they might use differently than if they were asked to use their own money), the money they are given (in our case, $12) usually isn't that large a sum to put on the line, and the students are drawn from a similar cultural pool. But they also allow us to have a lot of control and ensure that participants have a similar experience so that we can focus on the variables in question—in our case, honor and shame.

In our setup, participants could see a screen where all the donations were listed under a fake name for that participant, so that participants were anonymous to one another and to us. We ran three different treatments: shame, honor, and a control. In the shame condition, we told the students that after ten rounds, the two least generous players would be asked to come forward and write their real names on the board in front of their fellow group members under the heading "I donated least." For honor, the two most cooperative players were called forward and wrote their names under the heading "I donated most." Students knew the setup before the game began and could therefore avoid being the least or most generous if they chose to. We recruited each group of six players from the same class, and the experiments occurred early in the term, so that any reputation effects might last for the remainder of the semester. In the control condition, all players, including the most and least generous ones, remained anonymous for the entire game. (To me, the control tested the guilt condition, where participants made decisions based on internal judgments rather than possible reputation and what the group might think.)

After the experiment, we asked participants what their strategy had been when they decided to give or not in each round. Responses included: "My strategy was to donate as little as possible without being exposed as someone who contributed least." "I did not want to be one of the least generous players, so my only aim was to stay out of the bottom two, other than that I tried to maximize profit." "Towards the 5th–6th rounds, my trend of thought changed, and I started paying attention to the individual contributions to make sure I was not in the bottom two." (Not all decisions were sensible, which won't shock anyone who has spent time watching the Home Shopping Network. Students also reported strategies like "If I pulled out a coin and it was heads, I donated. If it was tails, I wouldn't donate" or said that they had given "only in even rounds, plus the lucky number seven.")

Other experiments had tested the effects of transparency—when all players are exposed to one another—

and showed that it enhanced cooperation, but in our experiment only a minority within the group was exposed. Yet our results showed that most participants responded to the threat of shame and tried to avoid it by being more generous. On average, students were 50 percent more cooperative in both the shame and honor conditions than in the control, where players remained anonymous. People were willing to pay to avoid shame—as well as willing to pay to achieve honor.[8]

But if transparency and honor also work, why write a book focused on shaming? Transparency is more democratic, but in the last chapter we will find some reasons that shaming policies can actually be more protective and effective. Meanwhile, honor is less painful and less awkward, and in our experiment, where giving to the public pool was optional, honor was a decent strategy that increased giving. But one of honor's shortcomings is precisely its optional nature—not everyone seeks it, while most of us seek to avoid the taint of shame, or losing face. Also, it's often the least cooperative people who cause us the most concern.

Consider taxes. Everyone benefits from universal social services provided by the tax system, and the vast majority of Americans pay their taxes. Imagine a list that attempted to honor people for paying their taxes—it would be absurd. We simply expect people to pay their taxes. But there is a small minority of high-income people who choose not to pay their taxes, and these people—specifically the ones who owe the greatest amounts in back taxes—are the issue. The State of California's website describes the cooperative dilemma: "When taxpayers do not pay their fair share it places an unfair burden on those who do. Closing the tax gap is in the best interest of all Californians." The tax gap is about $10 billion, which means $10 billion less in shared state services like schools, roads, and medical care. Everyone in California is a victim of tax transgressions. In the case of the federal government, there is a system of formal punishment to deal with people who do not pay taxes, which includes prison, but at the state level, formal punishment is functionally absent.

Since 2007, the State of California has annually published an online list of the top five hundred individual- and business-tax delinquents whose outstanding taxes for the previous fiscal year exceed $100,000. Note that because of this requirement, and perhaps contrary to our instincts, this shaming policy is more protective than transparency. A transparency policy for tax delinquents would expose all people who owe money to the government, no matter how little money they owe, while the shaming policy (which posts only the worst delinquents) is more protective, because it focuses the audience's attention on the people most responsible for the tax gap while at the same time protecting the poorest citizens. Errant taxpayers receive letters in advance of the list's publication (which happens twice per year), with the expectation that the threat of exposure will lead to payment, at least from some debtors. If they pay their back taxes, they can avoid being listed. The State of California has so far retrieved over $336 million—a sum that dwarfs the program's estimated $131,000 annual cost. More than twenty U.S. states have implemented similar programs on their sites.

Honor was not the motivation behind tax payments, nor was guilt. The threat of shame was what worked. But just because shaming can encourage cooperation or conformity doesn't mean people should be threatened with exposure at every decision. We also know that bribes can be effective at encouraging certain behaviors, and yet bribery is not encouraged. The threat of shame may be more effective than the actual experience. Like antibiotics, shaming works best when used sparingly. And also as with antibiotics, if shaming is abused, we might all end up as victims.

Obviously not all shame or shaming is good. Everyone agrees that it's wrong to steal, but we would not support the resurrection of tarring and feathering or a pillory in which a cabbage thief had to stand with cabbages on his head. Likewise, William F. Buckley's proposal in the 1980s that all people diagnosed with AIDS should be required to get a standard tattoo on their upper forearm and buttocks to prevent the spread of the disease was unacceptable. Even today, some judges require criminals to wear a T-shirt announcing their crime, and many people disagree with this punishment. In addition to shaming, these punishments each contain an aspect of humiliation, going beyond mere exposure to include a form of stigmatization. Scholars have argued that shaming that takes this form is dehumanizing and strips transgressors of their personal dignity.

While ideally the best way to avoid shame is to avoid being its subject to begin with, sometimes conforming to the group's desires is just too expensive, and sometimes it's just too late. Sometimes people just don't want to become the person the group wants them to be, and other times they cannot. The volatility and variability in how people react to social disapproval is part of shaming's liability. The threat of shame also isn't going to deter everyone—some people are shameless. But that's why we have forms of punishment that extend beyond reputation. If shaming were a perfect regulator of human behavior, we wouldn't have a long history of using ostracism, physical abuse, prisons, and even (in some countries) capital punishment. (Note, however, that at the state level none of these punishments are available for tax delinquents.)

The Stakes

Around the year 1800, with the Industrial Revolution under way, humans became something unprecedented: a globally transformative force. This new era has been called the Anthropocene, and it has many alarming features, most of which are expressed in the form of line graphs that look like Mount Everest sliced in half vertically: human population, biodiversity loss, atmospheric carbon dioxide emissions, water use, number of McDonald's restaurants—all steeply increasing. In several cases, like climate change, we have exceeded boundaries of pollution that are safe for humanity and are likely causing dangerous environmental change.

Midcentury fantasies that humanity would have by now colonized the deep sea and outer space seem quaintly delusional. By 2050, an estimated ten billion people will be eking out a life on earth. Despite the promises, not one of us lives in outer space or the deep sea. As the lone survivor of the *Homo* genus, we have a unique and occasionally overconfident view of ourselves. Not that our species isn't special: we have complex language, we often cooperate with individuals to whom we are not related, and we have colonized or had some impact on almost all of the planet. We might also be the only organism to reflect on our own mortality. We alone as a species are faced with determining the fate of much of the world's nonhuman life.

We need new rules in this new era, and for those, we are

going to need a lot of help from shame. We have traded in crude forms of shaming for more sophisticated, less offensive forms. When we're all in it together—as we are with the destruction of the ozone layer, the threat of a nuclear holocaust, or the spread of infectious diseases—and when it comes to dilemmas in which a few bad apples can spoil things for everyone else, shaming might become more acceptable as a means of social enforcement, because the audience is also a victim of the transgression and because we have very few other options of punishment.

Shame is not only a feeling. It's also a tool—a delicate and sometimes dangerous one—that we can put to use to help solve serious problems. Shaming is a nonviolent form of resistance that anyone can use, and, unlike guilt, it can be used to influence the way groups behave—shame can scale. But shaming requires the attention of the audience, and attention is a zero-sum game. So shaming must be used shrewdly to maximize effectiveness, and this book helps to define what "shrewdly" in this case means.

The right amount of shame has helped us to get along, to the extent that we have, and has coordinated social life to make it a little less painful, a little more dignified. It has reminded us that we are, in fact, in this together. If used wisely, the same attention to one another that promoted our own evolutionary success can perhaps keep us from failing the other species in life's fabric and, in the end, ourselves. But for good reasons, shaming has acquired a controversial reputation, and it is important to understand why.

2

Guilt's Ascendancy

> A healthy dose of guilt never hurt anybody. It's what civilization was built on, guilt. A highly underrated emotion.
>
> —BARACK OBAMA (remembering what his mother told him), *Dreams from My Father* (1995)

Edward Kloberg III was born in New York City in 1942, when cars still mingled with the occasional horse-drawn cart. He flunked out of Princeton University, eventually graduated from a small college in New Jersey, and then got his graduate degree in history at American University, in Washington D.C., where afterward he had a lackluster job in fund-raising. But like many Americans, Kloberg believed he deserved more than lackluster. Being a III might sound grand, but it wasn't grand enough.

While still in his twenties, Kloberg changed his name to Edward van Kloberg. He later modified the Dutch "van" to a German "von" and then adopted the title of baron. At age forty, the self-styled Baron Edward von Kloberg III started a public relations firm, and ten years later he had built up such a controversial list of clients that *Spy* magazine referred to him as "Washington's most shameless lobbyist."

Von Kloberg's motto was "Shame is for sissies." And von Kloberg was no sissy. He represented Saddam Hussein, Zaire's Mobutu Sese Seko, and the Myanmar regime. He defended the Honduran Apparel Manufacturers Association against charges of sexual abuse and child labor, even though he knew they had transgressed, reasoning that if lawyers could defend such clients, so could a public relations firm. If shame was for sissies, then the question remains: Why were von Kloberg's reputation-management services sought by so many powerful people?

Shame might have been important in our evolutionary history, but people like von Kloberg make us wonder if shame is still relevant today. Psychologists June Tangney and Ronda Dearing view shame as "a primitive emotion that likely served a more adaptive function in the distant past, among ancestors whose cognitive processes were less sophisticated in the context of a much simpler human society."[1] Is shame really a phenomenon that belongs only to the distant past?

The Rise of Guilt

Some of the first anthropologists to study shame, like Ruth Benedict, claimed it was more important to collectivist cultures, like those in Japan but also in China, Brazil, Greece, Iran, Russia, and South Korea, and less important to individualistic cultures in the West, where shame was supplanted by the self-punishment of guilt. Anthropologist Dan Fessler designed an experiment to test whether shame was less important in the West.[2] He asked a focus group in collectivist-minded Bengkulu, Indonesia, to make a list of the fifty-two most commonly discussed emotions and did the same with a focus group from individualistic Southern California. Then he asked eighty Indonesians and seventy-five Californians to rank the frequency of those fifty-two emotions in each of their societies. *Shame* was second on the Indonesian list, but it didn't make the top ten for Californians, not even close: it was listed as the forty-ninth most frequently discussed emotion in California, preceded by *grief* and followed by *contempt*. Another interesting difference was that Indonesians listed *afraid* in the top ten, while Californians included *bored* and *frustrated,* which I attribute at least partly to Los Angeles traffic. Meanwhile, Californians ranked *guilt* thirty-second, between *offended* and *disgusted,* while none of the top fifty emotions translated directly to *guilt* in Indonesia. In fact, in many Asian cultures, there is no word for guilt at all.

Guilt is believed, by and large, to be an emotional construct of the West. Guilt not only appears to be more widespread in the West, but it seems more prevalent today than it was in the past. The word for *guilt* does not appear in the Hebrew Old Testament. Shakespeare used the word *guilt* only 33 times, while he used *shame* 344 times.[3] We do not even know what guilt looks like: in a study of Wisconsin undergraduates who were given photos of people expressing different emotions, they could not recognize expressions of guilt, the way they could anger, disgust, fear, and even shame.

Yet Westerners report frequent feelings of guilt. In a study conducted in the Netherlands, people reported feeling guilt for about two hours out of each day.[4] Why has guilt ascended on the moral stage? Perhaps it became more important as we began to have more opportunities to be physically isolated from the group, since some argue that guilt is experienced in solitude, without reference to an audience. It would be difficult to feel a private emotion without the privacy in which to feel it. Compare the median amount of time spent alone (zero) for a member of the hunter-gatherer Yanomami tribe in the Amazon to current living arrangements in the United States, where 28 percent of households consist of just one person. Herant Katchadourian, an emeritus professor of psychiatry and human biology, believes that guilt (which he described as a more self-conscious form of shame) appeared alongside the human capacity to produce symbolic objects, like cave paintings (the oldest of which are from forty thousand years ago), which imply the existence of abstract thought and the beginnings of new belief systems.[5]

In addition to physical privacy and symbolic objects, the rise of Christianity and, later, philosophies of individualism are also credited with guilt's ascendency in the spectrum of Western emotions. In the late eighteenth and nineteenth centuries, the notion of the individual was boosted to cultural prominence with phenomena like Romantic literature, which emphasized self-growth and self-expression, the American and French revolutions, and the understanding of individuals as agents of political change. Individualism gained a foothold as a moral position and a political philosophy—perhaps nowhere as much as in America, which is consistently ranked among the most individualistic of all countries. "Trust thyself: every heart vibrates to that iron string," wrote Bostonian Ralph Waldo Emerson in his 1841 essay "Self-Reliance."

Yet one of the many paradoxes of American culture is that it purports to believe in the individual, while it is simultaneously obsessed with greatness—and what is greatness if not a judgment bestowed by, and a status relative to, one's peers?

Today, American culture also champions corporations, which, operating mainly from the motive of profit and growth, seem a distant cry from the complexities of an individual and concepts of self-reliance. The axiom of libertarianism—a limited role for the state—grew out of a sincere belief in the individual (and yet some of the most self-professing libertarians, the league from Silicon Valley and the tech industry at large, are tethered to large companies, technologies, and investments that are profitable only because of their social nature).

The point is simply that the paradigm of the individual, whether real or imagined, elbows out shame, which is an inherently social phenomenon and, in our culture of self-reliance, can seem quaint or anachronistic. A hyper-individualist and privacy-loving society is left, at least allegorically, with guilt as its primary hope for social control.

Another, more practical reason guilt has gained power is that guilt promises to make punishment less expensive. If an individual can use guilt to police himself, the group or the state doesn't have to. Guilt is the cheapest form of social enforcement, because the norm has been internalized and is self-enforcing. Evolutionary biologist Robert Trivers put it this way: "The emotion of guilt has been selected for in humans partly in order to motivate the cheater to compensate his misdeed and to behave reciprocally in the future, and thus to prevent the rupture of reciprocal relationships."[6]

This leads to another and more contentious reason

for guilt's ascendancy. Some people suppose that guilt is not just cheaper, but morally better. Guilt, unlike shame, takes an act, rather than the whole person, as its primary object. Since at least the eighteenth century, philosophers have also argued that guilt is a beneficial emotion that motivates people to admit responsibility for their wrong actions, to make amends and repair damaged relationships, unlike shame, which makes us want to hide or disappear. Some argue that guilt is less painful than shame, and more refined, although it's clear that some people can be painfully afflicted with guilt.

The State's Role in the Decline of Shaming

With the rise of church and state as powerful and pervasive institutions, human groups relinquished certain powers, such as physical punishment, to these authorities. Before the invention of prisons, many punishments had a particularly ugly and public history. Gruesome punishments like quartering, in which a man was chained to four horses and had his limbs torn off, were carried out for all to see. Human branding dates from at least the twelfth century, and slitting of the ear or nose were also permanent physical stigmas. The Greeks invented the very word *stigma* to refer to the symbols burned or cut into a transgressor to show he was a slave, criminal, or traitor. According to anthropologist Gustav Peebles, who studies monetary history, branding a debtor was the earliest form of a credit rating. In the town square, there were whippings, ducking stools, and the stocks. A dishonest baker might be sent to the pillory with dough on his head.

Then a lot of shaming punishments in the West disappeared. Philosopher Michel Foucault described the state's role in the disappearance of shaming punishments in his book *Discipline and Punish: The Birth of the Prison* (1975). The pillory was put to an end in France in 1789, and in England in 1837. Branding was abolished in France and England in the nineteenth century, as was the public exhibition of prisoners in France. New York was the first state to end public hangings, in 1835, and the other states eventually followed.[7] Prisons helped make punishment a more private affair.

Despite the widespread use of prison sentences, which themselves come with a certain amount of shame, and the decline in shaming punishments, shaming by the state continues in milder forms. In the United States, judges in some states require convicted thieves to publicly carry signs that broadcast their crime. In November 2012, a woman in Ohio who had been caught driving on the sidewalk to go around a morning school bus full of children was sentenced to standing for two hours on the sidewalk with a sign that read: ONLY AN IDIOT WOULD DRIVE ON THE SIDEWALK TO AVOID A SCHOOL BUS. In China, it is common for criminals to be paraded around wearing signs describing their offenses.

There are many strong arguments why the state should not use shame against criminals. Martha Nussbaum, political philosopher and one of the most ardent opponents of shaming punishments, argued against state shaming because she believes that a primary responsibility of the state is to protect human dignity, and shaming punishments run counter to this aim. One might counter that the transgressions themselves also undermine human dignity, but to Nussbaum the difference is the involvement of the public system of law. "The fact that the state is complicit in the shaming makes a large difference. People will continue to stigmatize other people, and criminals are bound to be among those stigmatized. For the state to participate in this humiliation, however, is profoundly subversive of the ideas of equality and dignity on which liberal society is based."[8] Other legal scholars argue that shaming cannot be effective in the mobile, anonymous, urban societies of today.

Law professor James Q. Whitman argued against shaming because it transfers the responsibility to punish from the state to the public. His objections to state shaming are grounded in the idea that the state is supposed to alleviate citizens of their impulses or obligations to punish, not to invite them to partake in more of it. "Shame sanctions are wrong in our society for the same reason that we feel they are wrong in China, or in the Afghanistan of the Taliban: They represent an unacceptable style of governance through their play on public psychology . . . Shame sanctions would be wrong even if they had no impact on the offender at all, for, no matter what, they would represent an improper partnership between the state and the crowd."[9] In contrast to Nussbaum's argument, Whitman discouraged shaming for the sake of the audience, not the transgressor.

In other words, people have chosen a government to represent them, even in times of punishment. When the government invites the public to take part in the punishment—by requiring convicted thieves to publicly carry signs that detail their crime, for instance—it is delegating part of its duty to a populace that can, in Whitman's words, be "fickle and uncontrolled." But as we'll see later, there might be instances in which shaming, even by the state, can be used for the benefit of the crowd in ways that the crowd also finds acceptable.

Is There Something like True Shamelessness?

Not only does shame rank low on the American emotional frequency, but its opposite—shamelessness—is often prominent (prompting books like Aaron James's *Assholes*). Ruth Benedict found that the Japanese, for example, had not expected Americans captured during World War II to show no shame in being prisoners of war, since that was unthinkable in their culture. In our experiments testing shame's effect on cooperation, many students wanted to avoid being singled out, but some students were not deterred. I was surprised by some of the brazen faces of the players who were singled out to stand in front of their peers for contributing nothing to the game. It would be easy to attribute the lack of shame shown in these cases to an individual's general state of shamelessness, but it's probably more complicated than that.

Shame is calibrated to social norms, which is why it is so dynamic and differs so much across cultures. In central Brazil, Bakairí Indians find eating in public incredibly shameful, while they find no shame in their nakedness. You might be ashamed if you hear your date did not tip the waiter at a nice restaurant in the United States, but you would not feel the same way in France, where tipping is not customary. In many cases, shame and shaming have gotten a bad rap, when in fact the shame might not be as problematic as the norm to which that shaming is calibrated. When doctors feel too ashamed to admit or to share their mistakes, the real culprit is not shame, but the norm that dictates that doctors cannot and do not make errors (driven largely, I imagine, by threats of litigation).

Part of shame's potency is due to our negativity bias—

the asymmetry in how our brains process negative versus positive information. The negative stuff sticks better. People sometimes believe negative gossip even if it's it clearly inaccurate. Negative opinions are also more contagious than positive ones,[10] and losing something (including, perhaps, one's reputation) feels particularly bad—worse than gaining that same something feels good.[11] For these reasons, parents repeat proverbs like "One falsehood spoils a thousand truths" or Ben Franklin's "It takes many good deeds to build a good reputation, and only one bad one to lose it."

Shame is more conspicuous in collectivist cultures in part because the norms are more commonly shared and widely enforced. In societies that value individualism and outliers, shame might be less apparent. But shame still exists everywhere. Many people believe that violent gang members are shameless, and certain gang members might be. But gangs, like any group, have their own brand of shame, calibrated to their own set of norms—often related to loyalty and snitching—and their own forms of punishment. Even among thieves, there is honor and there is shame—they are just provoked by a different set of standards.

Even Edward von Kloberg III felt shame. In 1998, he fell in love with a man who did not love him as much in return. His business started losing money, and there is nothing more shameful for a rich American than becoming poor. On top of that, von Kloberg got diabetes, skin cancer, and an inner-ear disease. Yet he refused to accept charity from his friends, or what he referred to as "welfare" treatment from hospitals. The same man who infamously declared that shame was for sissies became literally mortified with shame. In 2005, in a final performance consistent with his affinity for melodrama, sixty-three-year-old von Kloberg jumped to his death from a Roman castle.

Differences Between Guilt, Shame, and Embarrassment

Many psychologists agree that guilt is a private emotion, while embarrassment and shame are more public emotions, most often felt in the presence of other people. (Here we can all recognize that what makes us feel guilty is still informed by our community.) But exposure isn't the entire story; otherwise there would be no difference between shame and embarrassment.

As part of an experiment, students at UCLA were asked to give a five-minute speech, as well as complete a five-minute mental arithmetic test while annoying music played. A loud alarm went off if the student did not answer in time. One group completed the tasks alone, and the other, less fortunate group gave the speech and took the math test in front of two stony-faced peers. Subjects were asked afterward how they felt about themselves and also gave saliva samples four times during the experiment so that researchers could measure the stress hormone cortisol. Compared with the students who performed alone, the students who were observed by their peers reported more anxiety and had twice as much cortisol in their bloodstream.[12]

How do we know that the stressed-out students in this experiment were feeling shame and not another emotion, like fear, humiliation, embarrassment, or guilt? One way to tell the difference between emotions is by looking at emotional displays. We can watch just the face for the earliest emotions in life, like joy, sadness, fear, and anger, but for self-conscious emotions like shame, we must also watch the body.[13] In shame, the corners of the mouth are down-turned (sometimes the lower lip is tucked between the teeth), the eyes are lowered, the body collapses and tilts forward, and the shoulders fold in what seems like an attempt to hide. In some cultures, people cover their face with their hands. In Southeast Asia, sometimes people show shame by biting their tongue.[14] Embarrassment, on the other hand, often involves smiling, rubbing the face or other nervous touching, and the eyes wandering from side to side.

A display of shame is a reaction to exposure. It's not about what's going on inside, exactly, but about visible changes outside. These signals, like most others, are prone to deception—it's easy to fake a smile. The exception seems to be the blush. Other than by painting or powdering it on, a blush cannot be faked, which is why scientists refer to the blush as an honest signal. The blush is uniquely human and is one of our most nonthreatening displays.

A blush can be triggered by shame or embarrassment and interpreted as a sign of fault, submission, appeasement, or even arousal, depending on the circumstance. Undergraduates in the Netherlands read to themselves a series of vignettes and found that when the situation was morally ambiguous, such as running into someone who had said they were going to be somewhere else, they took that person's blushing as a sign of culpability. However, in instances where there was a clear-cut transgression, such as coming upon someone damaging your bicycle as they tried to remove their own from the rack, a blush from the transgressor was pacifying.[15]

"The blind do not escape," Darwin wrote in *The Expres-*

sions of Emotions in Man and Animals (1872). A reverend who had visited an asylum had informed Darwin that three children who were blind from birth were "great blushers."[16] Darwin's curiosity about blushing led him to speculate on the innateness of shame. It was already known that some behaviors, like smiling, were innate, while others, like the handshake, were learned. And Darwin's intuition was that emotional expressions that were innate were elevated in terms of evolutionary importance.

Darwin had the anecdote about the blind, but not the experiment to test it. At the 2004 Olympic Games, psychologists Jessica Tracy and David Matsumoto analyzed photographs of 111 judo athletes from around the world immediately after their matches. Winners tended to tilt their heads back, raise their arms, or expand their chests. Losers—the ones more likely to experience shame—tilted their heads down, hid their faces with their hands, slumped their shoulders, and caved their chests. To get at the question of innateness, the researchers also photographed Paralympians, fifty-three of whom were blind—at least twelve of them since birth—and they all showed similar postures, confirming that humans are born with an automatic shame display.[17] At some point in human prehistory, if not today, showing shame must have been a very useful thing.

A display of shame has the potential to deescalate conflict. Showing shame can hasten forgiveness. In a study involving a mock trial, a defendant accused of selling drugs who showed embarrassment and shame received two-thirds the prison sentence and was nominated for parole more than a year earlier than he was in other mock trials in which he showed neutral behavior or contempt.[18] This is probably also why lawyers advise clients not to succumb to the new, shameless trend of smiling in mug shots.

Another way to tell the difference between emotions like guilt, shame, and embarrassment is to ask why or when they are felt. U.S. undergraduates described circumstances that made them feel guilt, embarrassment, and shame. Analysis revealed that the students felt guilt over private failures like lying, neglecting friends or family members, breaking a diet, or cheating. They felt embarrassed when other people witnessed their minor mishaps, like physical accidents or lapses in memory, such as forgetting an acquaintance's name. For shame, the shortcomings were not only more public, but also more serious: poor grades, hurting people emotionally, and failing to meet someone else's expectations.[19]

Shame often extends to our conception of our whole self—the type of person we are—whereas embarrassment is associated with an involuntary, isolated incident. To differentiate between shame and embarrassment in children, researchers subjected five dozen four-year-olds to a series of uncomfortable events, like being complimented excessively, pointed at unexpectedly, persuaded to look at their reflection in a mirror, and asked to dance while the experimenter watched. They were also asked to complete, within an allotted time, four color-matching puzzles—two that were actually impossible to solve and two easy ones that the four-year-olds were guaranteed to finish because experimenters would postpone the alarm. The children gave saliva samples and were filmed from behind a one-way mirror. Later, researchers coded the children's facial expressions and body language for expressions like pride, embarrassment, and shame. The children showed embarrassment, which included face touching and side-glancing, when they were the center of attention, including while being complimented a lot and being asked to dance. But they showed higher cortisol levels and shame displays when they failed to complete the color puzzles in time.[20] A four-year-old feels embarrassed when some strange lady expresses how handsome he is, but he feels shame when he cannot put together the color puzzle he was told every other kid his age could solve.

With both embarrassment and shame, someone is watching, but in the case of shame, being seen matters more. You might feel embarrassed for unwittingly violating a norm of convention—if you are underdressed for an event or if someone sees you with toilet paper stuck to your shoe—but you feel ashamed for something like cheating on an exam, because you have failed according to a more visible social standard. When the UCLA experimenters tested the mental agility of undergraduates in front of their peers, the emotion was closer to shame than embarrassment, because mental agility says something about the whole self, something you can't explain away. Embarrassment is more easily forgotten, since it is linked to an isolated incident, while shame, linked to one's identity, has staying power.

Guilt's Successes

Shame resides deeper in human nature, but guilt should not be ignored as a powerful force. Guilt is claimed to be more civilized, less awkward, more likely to lead to reparation, and cheaper than shaming, which requires the group to pay attention to who is being singled out for what, and to care. Guilt obviously can work to motivate behavior, as it did when I found out that my family's tuna sandwiches were killing dolphins. Experiments show that individuals who receive feedback about their own performance often change their behavior, such as by decreasing their energy or water use, which some might say is the result of guilt. (Others might argue that it's the result of being monitored.) A recent study found that first-time blood donors were often motivated by external social pressures, like obligations to a group, but if donation continued, their self-image changed and they came to see themselves as donors—and this internalized view could then rely on guilt for self-reinforcement and regulation. Another study of volunteers for conservation activities found many were motivated by the desire to feel less guilty for environmental problems.[21]

In a study of 369,211 registered voters in Michigan, experimenters sent some voters a letter encouraging them to vote, and another group a letter encouraging them to vote while also indicating whether they had voted in the previous election (which is public information). Because subjects knew they were being watched by experimenters, it is impossible to say whether the results were purely the result of guilt, but the outcome does suggest that an internal voice was at work. Letters that encouraged people to vote led to a slight increase in voting (by 2 percent), but information about whether the individual had voted in the last election doubled the effect if that individual had voted in the previous election (a 4 percent increase) and tripled the effect if the individual had not (a 6 percent increase).[22] The voter's guilty conscience protested loudest when he failed in his civic duty, and incited the greatest behavioral change. Guilt, like all emotions, has its strengths, but it also has its limits when it comes to what it can accomplish, at what scale, and how quickly.

3

The Limits to Guilt

> My doctor says that I have a malformed public-duty
> gland and a natural deficiency in moral fiber, and
> that I am therefore excused from saving Universes.
>
> —DOUGLAS ADAMS, *The Hitchhiker's Guide to the Galaxy*
> (1979)

In 2007, Al Gore and the Intergovernmental Panel on
Climate Change (IPCC) won the Nobel Peace Prize for
their work on climate change, *Vanity Fair* launched its
annual green issue (which was terminated two years
later), and my friend Kate heard about upscale designer
Anya Hindmarch's release of canvas totes on which was
written I'M NOT A PLASTIC BAG. The Hindmarch bag was
just one example about how environmental concerns—in
this case the pollution caused by plastic bags—were being
expressed. The Hindmarch marketing machine had
already given bags in advance to celebrities in London to
create buzz, and Kate found a store that would be selling
the limited-edition bags for only £5.

Kate, one of my favorite misanthropes, asked her husband, Saran, if he would line up to get a canvas bag for me as "a relic of the environmental movement." Saran saw the assignment as "a challenge" and arrived at the store an hour before they gave out queue tickets. He was, conspicuously, one of the only men, one of the only non-whites, and one of the only people who stayed calm as the crowd of more than a hundred women, including a large contingent of preteen daughters, became more and more irritable. More than an hour later, a store manager gave him a ticket for the canvas-bag queue and told him to "be discreet." It became obvious that some people were not going to get tickets and, therefore, bags. Some in the line mentioned things like injustice. A woman who realized she would go home without a Hindmarch began to cry. At this point in Saran's recollection, Kate interjected, "The world is so pathetic."

Also pathetic: the canvas bags came wrapped in plastic. While I thanked Kate and Saran and kept my relic, others quickly sold theirs online for £200. The Hindmarches were every bit as intoxicating as Beanie Babies, cronuts, and gold iPhones. "Everyone felt really embarrassed afterward at how excited they had gotten," Kate said. "So there was this backlash. Plus every crappy merchandiser in London made a knockoff." An article in the *Observer* called it THE YEAR OUR HANDBAG HABIT GOT OUT OF CONTROL and reported that riot police were needed to contain the hysteria over the Hindmarch bags in Taiwan, where thirty people wound up in the hospital. In Hong Kong, the Hindmarches shut down a shopping mall. Until recently, this was not behavior associated with the environmental movement.

Green Guilt

One effect of our growing awareness of social and environmental problems has been a rise in guilt. Buy a toilet seat from Walmart and know that a forest in Russia was probably leveled on its behalf. Buy a diamond and worry that it financed a war in Africa. Eat shrimp sourced from a farm in Bangladesh and contribute to the labor abuses of women and children. Upgrade to a new iPad and the old e-waste goes to a landfill in Nigeria, where lead, cadmium, and mercury leach into the soil. "Tell your parents not to ruin the world you will live in" was one sentence flashed at the end of the 2006 film *An Inconvenient Truth*. Wildlife biologist George Schaller, in an interview with *Discover* magazine, said, "Obviously humans are evolution's greatest mistake."[1]

I asked Lynn O'Connor, a psychologist and an expert in survivor guilt, whether she had any patients concerned with the fact that humans are proliferating, while so many other species go extinct. "It's not a presenting symptom," she said. "But I had one patient who was obsessed with global warming. She was really worried about it, and the thing is, I think she was right to be worried." The American Psychological Association's task force on climate change, in its 2009 report, included accounts of "eco-anxiety," whose symptoms include panic attacks, loss of appetite, and sleeplessness.

Mothers ridden with environmental guilt reuse their children's bathwater. Shoppers refuse to buy blueberries imported from Chile due to the guilt they feel about the fuel used in shipping. A woman feels guilt over the wild habitat lost to cocoa plantations and refuses to buy chocolate, prompting her husband to say she took the joy out of his Almond Joy. A man who feels guilty about population growth insists on adopting if his wife wants more than one child.

The Voluntary Human Extinction Movement (VHEMT), which launched its website in 1996, is now explained in twenty-five languages. The movement promotes a simple motto: "May we live long and die out." Having children today is not only a choice, but a decision that comes with a certain amount of guilt, and members of VHEMT have found this guilt potent enough to snuff out arguably the most fundamental biological desire. People join the VHEMT for a reason that goes beyond wanting time alone to do crossword puzzles or go on European absinthe tours: because (according to the VHEMT website) "phasing out the human race by voluntarily ceasing to breed will allow Earth's biosphere to return to good health." While some people refuse to fly or have children, or commit to conserving water, the more obvious way that we have come to engage with green guilt and anxiety has been through consumption.

Modern Indulgences for Green Guilt

Just as the devout purchased guilt-alleviating papal indulgences in the Middle Ages, guilt-ridden consumers today buy dolphin-safe tuna, compact fluorescent lightbulbs, hybrid cars, and Ethos Water (one of the many well-meaning companies to proselytize at the TED conference—later bought by Starbucks for $8 million). Carbon offsets share the greatest kinship with indulgences and are often framed in religious terms. An article in *The New Scientist* began, "If you must fly or indulge in other carbon-intensive activities, carbon offsets now promise redemption." *The Economist* ran the article "Carbon Off-sets: Sins of Emission." Peter Schweizer, co-editor of the book *Landmark Speeches of the American Conservative Movement* (2007), asked, Why stop with carbon? He suggested (in humor, I think) several other offset options, including one for adultery, where funds raised would be donated to Focus on the Family and make the offenders "adultery neutral."[2]

In addition to religious language, there is a lot of framing of offsets around dieting. Environmental groups have promoted a "low-carbon diet" and "carbon-counting" to encourage people to look at their carbon output the way they do their caloric intake. Weight loss is a ludicrous way to frame the problem of climate change. An individual might be able to control her figure, but the same cannot be said for an individual's power over the atmosphere, or any other environmental problem. Guilt-free products are also almost always more expensive, because, as the free-market logic goes, the costs are internalized rather than externalized to the environment. So just as the rich could buy their way out of penance, the rich can now presume to buy their way out of environmental destruction and its associated guilt.

Labels, Labels, Labels

The dolphin-safe logo of 1990, which eased the consciences of schoolchildren, including me, arose in a context of free-market ideology in which individual consumers, not government oversight over large-scale producers, were idealized as responsible for how things are produced. (In truth, regulation was key to decreasing dolphin deaths in tuna-fishing gears.) Lots of other eco-labels now exist,

from "cage free" to "free range" to "grass fed" to "all natural." The framing for these labels is useful, because they show us the default production practices, like factory farming and synthetic additives.

Many of these labels are misleading both consumers and conservation funders and leading to undesired complacency. One label I know well is the Marine Stewardship Council (MSC), created in 1997 to certify sustainable fisheries. The MSC label is meant to distinguish seafood caught with good fishing practices from that caught with bad fishing practices, although the seafood itself might be similar. In this way, the MSC eco-label is very different from the "organic" eco-label, which differentiates the product itself as pesticide-free, not only that product's means of production.

Today the MSC logo is on more than 180 fisheries but has failed to demonstrate improvements on the water. In protest, environmental groups have paid hundreds of thousands of dollars to formally object to MSC certifications (nineteen total objections to certification so far; only one has been upheld), claiming the MSC's principles for sustainable fishing are too lenient and allow for overly generous interpretation by the third parties that actually do the certification. Groups like the National Environmental Trust objected to the certification of the Gulf of Alaska pollock fishery and pointed out that the fishery does not comply with endangered species laws to protect the food supply of Steller's sea lions, and "respect for laws" is one thing the MSC requires for certification. When challenged, the MSC responded that "respect for laws" is "different to compliance" and "does not require that a fishery management system be in perfect minute-to-minute compliance with every single piece of substantive or procedural law that may govern a fishery." The MSC has replaced its intention to protect ocean species with a word game.

The number of eco-labels continues to increase, even though no studies suggest that eco-certifying fish has led to more fish in the sea, or that certifying wood has increased forest cover. The organic food industry is worth $30 billion but represents only 4 percent of the food market. From 2000 to 2007, the United States did decrease its pesticide use by 8 percent, which sounds pretty good until you realize that this means a reduction from 1.2 billion pounds to 1.1 billion pounds. These labels alone are not getting us where we want to go. And yet, in 2011, researchers suggested in the journal *Nature Climate Change* that it was "time to try carbon labeling" and that "the existence of shortcomings [in labeling and certification] does not obviate the value of such a program."[3] What does?

Products labeled "all natural" in the United States are so prolific, and the label so meaningless (and lacking in any regulation), that a nonprofit coalition is now working to eliminate it altogether. (Because what does the label that I've seen on some sugary beverages "naturally flavored with other natural flavors" actually mean?) But today the efforts of governments, university scientists, and nonprofits to point out the problems with labels are seldom a match for business interests.

Recent investigations suggest that the relatively successful "organic" label is also, at least in the United States, headed for trouble. Signs of regulatory capture are evident, with large companies like General Mills, Campbell Soup Company, and Whole Foods Market on the standards board, which attempted, in one case, to add a synthetic herbicide to the list of what can count as organic.[4]

Walmart has been accused of peddling fake organic goods on more than one occasion.[5] In 2007, the retailer was exposed for using in-store signage to mislabel foods as organic at dozens of stores in five states, and in 2011 for mislabeling conventional pork as organic in China, where the company has more than 390 stores.[6] And while this is not forgery per se, organic sea salt commands a higher price, despite questions about what organic in this case could mean.

Okay, but even if green consumerism is irritating or demoralizing, it's still arguably harmless, right? In his 2007 article "Buying Into the Green Movement," published in the *New York Times* Fashion & Style section, Alex Williams wrote that environmentalists are not deluding themselves that buying eco-friendly products actually makes a difference. Green purchases are merely "a good first step" and "a beginning, not an end point."

Some evidence from work on moral licensing disagrees with this assumption that buying green is a good first step. People who buy eco-products can apparently more easily justify subsequent greed, lying, and stealing. A 2009 study showed that participants who were exposed to green products in a computer-simulated grocery store acted more generously in experiments that followed, but that participants who actually purchased green products over conventional ones then behaved more selfishly.[7] A 2013 study confirmed suspicions about slacktivism when research showed that people who undertook token behaviors to present a positive image in front of others—things like signing a petition or wearing a bracelet or "liking" a cause—were less likely to engage with the cause in a meaningful way later than others who made token gestures that were private.[8] This research suggests that linking "green" to conspicuous consumption might be a distraction and lead to less engagement later on. If this is true, we should not be encouraged to engage with our guilt as disenfranchised consumers, capable of making a change only through our purchases, and instead encouraged to engage as citizens. Markets might even undermine norms for more serious environmental behavior. In some cases, as has been noted in Western Australia, eco-labeling fisheries may even be giving fishing interests leverage against establishing marine protected areas, where fishing would be prohibited or more heavily regulated, on the grounds that protection is not needed if the fisheries in those areas are already labeled eco-friendly.[9] The market for green products might sedate our guilt without providing the larger, serious outcomes we really desire.

When consumers feel guilty about pesticides or unfair trade, they can avoid that guilt by buying organic foods and fair-trade products—and they do, in places like Whole Foods supermarket, dubbed by Nick Paumgarten, in his 2010 *New Yorker* article, "Holy Foods, the commercial embodiment of environmental and nutritional pieties." The Whole Foods business philosophy is perfectly consistent with its success: labeling should be voluntary, and consumers should have the freedom to choose (or not) healthy or eco-friendly products, and Whole Foods offers both, but many more products of the organic variety than conventional grocery stores. Voluntary standards, eco-labels, and consumer choice are what give Whole Foods its business edge and make it the place for eco-conscious consumers to shop. If every grocery store were required to sell organic or sustainable foods, Whole Foods would have to find some other way to distinguish itself.

So instead, most consumers continue buying the same

old stuff. For most social and environmental labels, only the portion of the industry that wants to cater to consumers with guilt-prone consciences needs to change. The rest of the industry can continue to use pesticides, or unfair trade, or destructive fishing gear—and can sell those products at lower prices. The next steps—rules to change an entire industry—are missing. This is all implicitly part of the plan, because the main incentive for producers to do the right thing—like grow organic foods or fish in sustainable ways—is a higher price for their product. To get a price premium, those products have to be the exception and not the rule, which means the market could ease the consciences of a few consumers but avoids making any imposed, long-lasting changes to the industry. This we can call guiltwashing—the deceptive use of guilt and guilt-alleviation techniques to engage people as consumers and convince them they are making a difference.

By contrast, consider the case of the growing hole in the ozone layer. In 1974, Sherwood Rowland and Mario Molina, two chemists at the University of California, Irvine, related the use of chlorofluorocarbons (CFCs) to the depletion of the atmosphere. (They were awarded, alongside Paul Crutzen, the 1995 Nobel Prize in Chemistry for their finding.) Destruction of the ozone layer did not slow because a handful (or even a majority) of consumers who felt guilty about the growing ozone hole began buying products that were CFC free. The CFC ban was implemented regionally three years after Rowland and Molina's discovery and then, globally, with the 1987 Montreal Protocol. "You cannot solve these problems with voluntary action, because most people will not volunteer," Molina told me. "It has to become government policy."

Misdirected Guilt

Another reason green guilt is so ineffective is simply that it's felt over the wrong things. When a list of the top twenty-five steps you can take for the environment includes "use rechargeable batteries," we should all pause—at least twenty-five other things are way more important than which batteries we buy. The very first recommendation at the end of the 2006 climate change documentary *An Inconvenient Truth* is "Buy energy efficient appliances and light bulbs." Yet household lighting accounts for only 2 percent of total U.S. carbon emissions and 6 percent of household energy use (excluding diet).

But the lightbulb recommendation stuck. When a 2009 survey of 505 subjects from seven U.S. cities asked what was the single most effective thing they could do to conserve energy in their lives, the most popular response was "Turn off the lights."[10] Almost 20 percent of respondents mentioned lighting, compared with 13 percent who said "Drive less," even though personal vehicles account for the biggest slice of U.S. household carbon emissions (nearly 40 percent) and 15 percent of overall U.S. emissions, not to mention that driving cars consumes more than six times more household energy than lighting. It would have been much better to talk about cars than lightbulbs.

But Americans aren't big on hearing about the big environmental impacts of cars. In a *Washington Post* article from 2010, the chief executive of AutoNation, the largest car dealership in the United States, said, " 'You have about 5 percent of the market that is green and committed to fuel efficiency, but the other 95 percent will give up an extra 5 mpg in fuel economy for a better cup holder.' "[11] Eco-markets did diddly-squat for average fuel standards in the United States, which was flatlined between 1985 and 2005. When the Obama administration signed legislation in 2012 that required automakers to make new cars and trucks with almost double the efficiency by 2025, consumer demand had nothing to do with it.

Americans don't own the patent on misdirected guilt. The British government's £22 million "Are You Doing Your Bit?" campaign, beginning in 1999, became preoccupied with energy savings by filling teakettles with just the right amount of water. In 2007, George Marshall, who writes about the psychology of climate change denial, pointed to the inanity in an op-ed in the *Guardian:* "A return flight to Australia [from the UK] will have the same climate-change impact as 730,000 plastic bags or 176,000 overfilled kettles."

In 2008, David MacKay, a physics professor at the University of Cambridge (whose first book was *Information Theory, Inference, and Learning Algorithms*), published *Sustainable Energy—Without the Hot Air*. The next year, when I met up with him at Cambridge's Darwin College, he said "the rubbish claims" and "the lack of numeracy" were what inspired him to write it. In the book, he showed that the frequent recommendation in the UK to unplug mobile phone chargers was a distraction, because this individual action would result in the same energy savings as not driving the average car for *one second*. More important, MacKay said it is not individual action that is needed, but legislation. "Government controls so many things," he said. "They determine the kind of cars we drive, the kind of buildings we build. Government determines the amount of renewable energy produced as a percentage of total. In the UK, we have become depressed and useless. And we *are* useless. We're allegedly a service economy, but who are we serving?" MacKay's work was a reminder that many green recommendations were not only off-base numerically, but they were wrong philosophically. So many campaigns (even ones funded by government) ask that citizens engage primarily as consumers.

Green Cynicism

Green consumerism's biggest crime is not its earnestness, but its delusion. We have been asked to engage with some very serious problems in some seriously silly ways, and deep down we know they are bogus. We instinctively know that engaging only as shoppers cannot work.

So it's no wonder that certain aspects of environmentalism have become targets for irony. In the *South Park* episode "Smug Alert!" one of the show's characters buys a "Toyonda Pious" and then, realizing how "backward and unsophisticated" his fellow townsfolk are, decides to move his family to San Francisco. Mick Stevens's *New Yorker* cartoon depicted a car dealer talking to a couple in a hybrid-car showroom: "It runs on a conventional gasoline-powered engine until it senses guilt, at which point it switches over to battery power." Barbara Smaller's *New Yorker* cartoon had a man asking a waiter, "Which of tonight's specials is the most sanctimonious?" Green consumerism was doomed to become a point of satire because it began to take itself too seriously. It began to look at itself as not only *a* solution, but *the* solution.

Each generation has its own brand of cynicism, and cynicism, like shame or guilt, does not suddenly appear or disappear, but is faithfully calibrated to the social dilemmas of its day. So while we have innumerable student groups trying to ban bottled water at their universities, we have *The Onion* selling water bottles that say MY OTHER WATER BOTTLE IS 30,000 STYROFOAM CUPS. Against the swaths of vegetarian restaurants and cookbooks, there is a pushback that includes the rise of bacon—"a baker's new best friend."

The Limits to Green Guilt

The same green guilt that might have mobilized a minority to activism has been co-opted by industry and used as a marketing tool to distract that same minority with the hollow activity of consumption. Writing about it now, I feel disappointed by the way I engaged on the dolphin-tuna issue, even though I was only nine years old. My take-home message from *50 Simple Things Kids Can Do to Save the Earth* left me more self-obsessed (measuring how much water we were losing from our leaky taps at home) rather than actively engaged with more systemic issues (agriculture accounts for the majority of water use). Not that guilt isn't motivating—it can, in some ways, be a healthy response to many of our problems. But the flaw comes when guilt is misguided and we find relief in shopping rather than activism, or when guilt over collective problems is used to improve oneself rather than to strategically consider the collective whole.

A handful of guilty consumers buying this or that was not what motivated car companies to increase fuel standards or what motivated Walmart to give its employees health insurance. It was not what got women the right to vote. Guilt-ridden consumers were not what convinced companies to stop the production of ozone-destroying chemicals.

Guilt also can be inadequate on its own terms because it

is inherently individualistic, and in today's world we often need changes at a higher level. In his book *The Corporation* (2003), Joel Bakan described how publicly traded corporations, with their exclusive focus on profit, behave like psychopaths. Corporations are arranged to be immune to guilt, because an employee's conscience cannot override the pursuit of profit. But corporations are also groups of individuals, the vast majority of whom are hardwired normally to operate with a set of morals and the full range of human emotions. So how is it permissible that individuals can get together and, as a corporation, act like profit-seeking psychopaths, but then return home and act according to their usual moral conscience? One reason, which Joel Bakan described, is the principle of limited liability, which "allowed investors to escape unscathed from their companies' failures" and undermined personal moral responsibility of individuals within a company. This is a feature of not just corporations but of national governments and militaries, too.

Yet small changes made by big institutions can make a serious difference—whereas small changes made by individual consumers cannot. Chevron's emissions in 2010 alone account for eleven times the emissions of all of the U.S. household lighting combined. Getting one single company to reduce its emissions by just 10 percent has a greater impact than getting every single American to agree to live in the dark. But, unlike individuals, Chevron cannot be motivated to change through guilt.

Recall anthropologist Margaret Mead's ubiquitous saying: "Never doubt that a small group of thoughtful, committed citizens can change the world. Indeed, it is the only thing that ever has." Mead spoke of citizens, not producers or consumers, although it's useful to consider her idea in market terms. While the market behavior of a few conscious people cannot solve a collective-risk problem, like overfishing, wildlife protection, or climate change, a relatively small group of producers or consumers *can* create a collective-risk dilemma. There can be an asymmetry in the impact of supply or demand, and a small group of committed people in the market can indeed change the world, but for the worse. This is the topic of the next chapter, and it is yet another reason why, when it comes to serious problems, we likely cannot rely on guilt alone.

4

Bad Apples

> The human species was given dominion over the earth and took the opportunity to exterminate other species and warm the atmosphere and generally ruin things in its own image, but it paid this price for its privileges: that the finite and specific animal body of this species contained a brain capable of conceiving the infinite and wishing to be infinite itself.
>
> —JONATHAN FRANZEN, *The Corrections* (2002)

"All animals weren't created equal," biologist Bob Paine told me, paraphrasing his most famous idea. Paine is a towering man and had a height chart in his University of Washington lab to show that he could literally oversee his students. A line marked –GOD– was above Paine's own line, at six foot six. (Chris Harley, one of Paine's last and tallest graduate students, told me that he dreaded the moment when he would be measured. Harley was indeed taller than Paine but was relieved that he measured shorter than God.)

As an experimental ecologist, Paine likes to play God in the intertidal—the part of the coast that is above water at low tide and below water at high tide—where competition for real estate is stiff. He manipulated different variables to study the relationships between animals and their environment, although he said the work became less fun after sea otters repopulated his study site and displaced him as top predator. He has spent nearly fifty years as a vertebrate among invertebrates along the Washington State coast. "I've always thought of myself as a wader, due to my size," said Paine.

In the late 1960s, Paine designed an experiment starring a five-legged carnivore. He visited his sites at Mukkaw Bay every two weeks during the summer and tossed away any purple sea stars ("starfish" is inaccurate, since they are invertebrates, not fish). By removing the purple carnivores, he found he caused an overabundance of their prey—mussels—and a sharp decline in the rest of the animals that were usually present. Without sea stars, mussels outcompeted sponges. No sponges, no nudibranchs. Anemones also were starved out, because they eat the animals that sea stars dislodge. Also no more sea moss, acorn barnacles, black turban snails, or urchins.[1]

The purple sea stars had disproportionate influence relative to their abundance, so Paine called them a "keystone species"—which has become, in conservation terms, a category of species especially important to protect. But in the most basic sense, being keystone also means having more influence on community structure than other species. Could the keystone concept have some relevance to human consumers? Paine considered my question. His eyes looked up, and his lips turned down into a "Why not?" Then he said, "Humans are a vastly nastier customer."

Statistics of Deadly Quarrels[2]

One species, *Homo sapiens,* has had a remarkable influence on global ecology. The impacts are so numerous and severe that we now live in a geologic epoch informally named after our power: the Anthropocene. (The official vote to accept the name takes place in 2016.) One of the Anthropocene's many characteristics is that, under our care, the extinction rate is now likely one thousand times higher than the average extinction rate over all of the earth's history, leading scientists to ask if we're in the middle of a sixth mass extinction (which is when more than three-quarters of species are lost in a geologically short time span—the last mass extinction included the disappearance of non-avian dinosaurs, sixty-five million years ago, and had nothing to do with humans).[3]

At the species level, the fragility of existence is not a long-standing human concern. The phenomenon of extinction was understood, as far as I can tell, at roughly the same time that evidence for the existence of dinosaurs was discovered, around two centuries ago—although I imagine some prehistoric humans must have wondered what had become of the woolly mammoth or the giant sloth. In 1824, shortly after the French zoologist Georges Cuvier recognized and popularized the fact that certain species were gone forever, the English paleontologist William Buckland described the first dinosaur fossil. The word *dinosaur* didn't appear for another eighteen years, when Richard Owen, scientist, museum curator, and opponent to Darwin's theory of natural selection, coined the term.

More than 90 percent of life that ever existed on the earth is now extinct, but it's not that surprising that humans didn't notice extinctions sooner, since most of them predate humanity—like when North America bumped into South America, around three and a half million years ago, and a substantial number of South American fauna went extinct after creatures from the north invaded (referred to as the Great American Biotic Interchange). Today, humans have more impact than the fusion of continents.

In many cases, a single member of *Homo sapiens* looked the last individual in the eyes before extinguishing the species. Goodbye, dear Steller's sea cow (once the world's largest sea vegetarian; last seen by a Russian harpooner in 1768), dodo (clubbed to extinction on the island of Mauritius in the middle of the seventeenth century), and Labrador duck (last one shot on Long Island on December 12, 1878). Goodbye forever, passenger pigeon (the last one died September 1, 1914, in the Cincinnati Zoo) and Tasmanian tiger (the last one died of neglect in the Hobart Zoo in 1936). Goodbye, Yangtze River dolphin (last seen in 2005), western black rhino (declared officially extinct in 2011), and, with Lonesome George's death in 2012, all Pinta Island giant tortoises (RIP). Since the sixteenth century, at least 870 known species have gone officially and, so far, irreversibly extinct.

At least another 17,000 species (which includes one-quarter of all mammals) are under threat of extinction, and for the moment are filed under the quieter "ecological extinction" category, where the species still breathes but there are no longer enough individuals to shape their surroundings. Mountain gorillas cling to small patches of a highland forest in central Africa (880 remaining), yellow-tailed woolly monkeys to the cloud forests in the Andes (250 or fewer), Ethiopian wolves to the cliffs on a Rift

~EXTINCT~

~EXTINCT~

~EXTINCT~

~EXTINCT~

~EXTINCT~

~EXTINCT~

~EXTINCT~

~EXTINCT~

Valley mountain (442), and a small number of northern right whales swim along North America's Atlantic coast (350 left). Some species are represented by fewer than 100 individuals, like the 55 Maui dolphins in New Zealand and the 6 remaining northern white rhinos (all in captivity). These species are not officially gone, but their numbers make us nervous and depressed.

Ruining It for the Rest of Us

In his famous 1968 essay, Garrett Hardin used the metaphor of "the tragedy of the commons" to describe the temptation to overgraze cattle on a shared pasture and therefore ruin it for everyone else. Because the benefits of adding a cow to the pasture are individualized but the costs are shared, "the rational herdsman concludes that the only sensible course for him to pursue is to add another animal to his herd. And another; and another . . . this is the conclusion reached by each and every rational herdsman sharing the commons," wrote Hardin.

For some shared dilemmas, like in Hardin's pasture, everyone is to blame. Consider noise levels in restaurants and nightclubs around Manhattan, such as the Biergarten at the Standard Hotel, which can be louder than the eighty-four-decibel C train. Who is to blame? Each patron contributes more or less the same amount to the cacophony, just as, in Hardin's commons, each herdsman contributes to overgrazing.

But in many other cooperative dilemmas, not everyone contributes equally. The problem is not "each and every herdsman," but one of a relatively few uncooperative herdsmen who can ruin the commons for everyone else. In the economics and business-organization literature, people who consistently destabilize cooperation are called "bad apples."

If we consider the group to be the remarkable array of species alive on earth, and cooperation to represent these species' survival, then humanity is the bad apple. But just as humanity stands out among the rest of earth's inhabitants as a particularly destructive force, certain humans stand out from humanity as particularly problematic. Although the Anthropocene implicates all of humanity, not all humans consume or pollute equally.

Consider anthropogenic climate change. Developed countries, including the United States, caused the extreme growth in atmospheric carbon dioxide concentration in the second half of the twentieth century. When carbon dioxide emissions are presented by country, the United States and China stand out as top contributors (and about twenty countries are responsible for 75 percent of global emissions). More specifically, the wealthiest 10 percent of humanity contributes an estimated 50 percent of global fossil fuel carbon dioxide emissions.

A 2013 research paper showed that just ninety corporations (some of them state-owned) are responsible for nearly two-thirds of historic carbon dioxide and methane emissions;[4] this reminds us that we don't equally share the blame for greenhouse gas emissions. *The Onion* spoofed, "New Report Finds Climate Change Caused by 7 Billion Key Individuals." This framing back to individual consumers is particularly untrue and unhelpful. Untrue because, of the seven billion individuals, the rich have had a much greater impact than the poor. Unhelpful because, even if all individuals were equally responsible for emissions (and they are not), corporations whose profits rely on extraction of fossil fuels have helped lock us into a system of fossil fuels. The oil-and-gas industry has lobbied the U.S. government (and others) against any policies that would decrease revenues—removing subsidies, enacting a carbon tax, funding renewable energy, or ratifying the Kyoto Protocol. Chevron—responsible for 3.5 percent of global carbon dioxide and methane emissions over the past century and a half—gave $1.1 million to U.S. political groups in 2008 alone, 75 percent of which went to Republicans, many of whom publicly and politically deny anthropogenic climate change.

In *Merchants of Doubt* (2010), Naomi Oreskes and Erik Conway described how "small numbers of people can have large, negative impacts, especially if they are organized, determined, and have access to power."[5] This particular brand of bad apple they referred to were scientists—mostly physicists—who had government positions during the Cold War and therefore had gained prestige and access to power. According to Oreskes and Conway, these scientists were motivated by a strict libertarian ideology and helped to fabricate scientific uncertainty around the linkages between smoking and cancer, coal use and acid rain, and fossil fuel use and climate change. These experts had a disproportionate impact on the debates relative to their number.

Patterns of asymmetry and disproportionate impact are everywhere. The United States has 5 percent of the world's population but holds 25 percent of its prisoners. The Lord's Resistance Army, in northern Uganda, apparently has only about 250 fighters but has managed to displace about 440,000 people. Violent street clashes that erupt after political or sporting events are normally the result of violence in 10 percent or less of the crowd.[6] An American eats 270 pounds of meat on average per year, while a Bangladeshi eats four.

In the health care business, certain patients also have a disproportionate effect on the system. A doctor in Camden, New Jersey, examined five years of data and found that just 1 percent of the 100,000 people who used the city's medical system accounted for 30 percent of its costs. (A single patient had 324 admissions during that time, and the most expensive patient cost $3.5 million.)[7] The overuse and improper use of antibiotics are leading to drug-resistant strains of diseases that could affect us all. Extensively drug-resistant tuberculosis (XDR-TB) is now a serious global problem, because a relatively small number of people did not take their full course of medications. People who forgo vaccinations for their children risk having them become carriers or victims of infectious diseases. While we are close to a global eradication of polio, the disease continues to be a threat because it is still endemic in Afghanistan, Pakistan, and Nigeria, where some religious groups are suspicious about the vaccination. (Gunmen killed nine polio vaccinators at two health centers in northern Nigeria in early 2013.)

Whether it's fossil fuel emissions, a quiet library, or the spread of infectious diseases, a relative few bad apples can spoil it for the whole bunch. But one could argue that the reverse is also true. There are bad apples who defy moral codes and corrupt cooperation, but there are good apples who mobilize and encourage good behavior. But, especially when it comes to cooperation, some bad apples are so rotten that they can undermine cooperation to such a degree that no other apple, no matter how good, can compensate.

Bad Apples and Rare Species

There are some dilemmas that can tolerate very few bad apples. Rare and endangered species are one example where demand from a very small pocket of people can wipe out an entire ilk of organisms. Today, the declines in both resident and migratory birds in Europe can be attributed mostly to residents of the Republic of Malta (the smallest population in the EU), Cyprus, and Italy, who like to hunt birds. The birds, which are often simply stopping through on their thousands-of-miles journey, are shot, trapped, and sometimes fried and eaten.[8] A relatively small group of hobby collectors, concentrated mostly in Japan, buys an estimated fifteen million stag beetles each year, sometimes shelling out $5,000 for a single individual.[9] Relatively small groups of people in France, the United States, Belgium, and Luxembourg import huge numbers of frog legs from Asia.[10]

What is the appeal behind bad-apple behavior? In 2007, a team of scientists led by conservation biologist Franck Courchamp went into an event at a Parisian luxury hotel and asked three hundred people or so if they would prefer caviar from a "rare" or "common" species of sturgeon, or if they had no preference. The French are serious caviar consumers and, combined with a few other markets, have driven populations of sturgeon around the world to the brink of collapse. (Turns out that killing a fish to eat its eggs is not a model for sustainability.) Before even tasting the caviar, 57 percent said they would prefer the rare one, and the rest did not express a preference; no one expressed wanting the common species. After the taste test, 70 percent said they preferred the rare species. But here's the wrinkle: both samples were actually caviar from farmed sturgeon. The eggs were the same. The heightened desire to consume one over the other was because one type was labeled as rare.

Luxury hotels are one thing, but what about more ordinary consumers? Courchamp and his team found almost identical results when they ran the experiment in three big suburban supermarkets.[11] They also ran a series of experiments at the oldest zoo in France and found that people were more willing to climb stairs, to get wet walking through a sprinkler, and to pay more to see a rare species than a common one. Zoo goers also stole more seeds from a display when they were labeled "rare" rather than "common."[12]

Wildlife collection is one of the world's biggest black markets. Bird watchers in Britain also value rarity and were eager to share unusual bird sightings on the Internet, but egg collectors would use this information to rob the nests of protected species. Just describing a new species can increase demand from collectors, as researchers learned after describing a turtle in Indonesia and a gecko in southeastern China.[13] Archaeologists face a similar problem, because simply describing sites can lead to increased pothunters and poaching.[14]

Rare goods make sense from an investment standpoint, which is why there are markets for art, antiques, rare books, rare coins, and rare fossils. French artist Bernar Venet has been known to buy back some of his early works to destroy them as a way to increase his market value and his reputation. A rumor has surfaced that rhino horn collectors are up to Venet's same tricks, allegedly sending in poachers to kill wild rhinos to increase the value of their already collected horns. In wildlife markets, creating rarity obviously has much more sinister ramifications. Even if just a relative few wildlife collectors value rarity, those individuals can be enough to cause anxiety for anyone worried about the fate of rare species.

The Science of Bad Apples

Normally, we can survive without thinking too much about distribution curves. But the nature of collective-risk dilemmas—problems we face together and all stand to suffer from—explains why we find some people's behavior more interesting than others. Consider the rise of vandalism in U.S. national parks or the problems with prescription drugs. We are not interested in people who do *not* graffiti a 150-year-old giant cactus or in doctors who are cautious in writing prescriptions. At the library, it is the people who fail to return the books who worry us, because they compromise the system for everyone else.

It seemed clear that our tendency in a collective problem is to be most concerned with the least cooperative people, but when I mentioned this to mathematical biologist Christoph Hauert, he asked about evidence. So we decided to gather some. In this experiment, subjects played a collective-risk game framed around climate change. Each person participated in a group of six, and each person was given a $20 endowment. As a group, the six players had to come up with a $60 donation to a "climate fund" over the course of ten rounds or risk losing all of their endowment with a 90 percent probability. If they succeeded in reaching the $60 goal, they got to keep whatever was leftover. This is a variation on the trusty public goods experiment, but with a threshold. The egalitarian outcome would be that each of the six players donated half of their endowment ($10 of the $20) to reach the

$60 goal, but this can be a difficult thing to accomplish, because the majority of players need to cooperate to achieve the threshold, and no single individual can ensure the group's success.

We ran twenty different games, and after each one we asked the participants which player's identity they would want to know most if they could—all 120 players were anonymous and would remain so, but each player had a pseudonym for the game, so we were able to ask the question hypothetically. The least cooperative player was by far the first choice, and this was true whether the group had succeeded at cooperating and reaching the $60 goal or not. People were most interested in the bad apple.

Other experiments have shown that free-riding causes strong negative emotions in other group members. After having people play a real public goods game, scientists then asked participants about their anger and annoyance toward a hypothetical free-rider in a hypothetical four-player game. The intensity of anger depended on the intensity of free-riding—in other words, how far the free-rider fell from the norm. Participants reported much more anger and annoyance when the free-rider supposedly gave two francs and the three other players gave fourteen, sixteen, and eighteen francs than they did when the free-rider gave two francs and the three other group members gave three, five, and seven francs.[15] In both cases, the free-rider gave two francs, but it made people angrier when those two francs seemed stingier. The degree of "bad" relative to the group matters when it comes to bad apples.

Another cooperation experiment found that free-riding had stronger negative impacts on group cooperation if the free-riding emanated from one or two individuals than if the same amount of free-riding was diffused among many players.[16] It is not defection itself that's upsetting, but the repeat defection by the same one or two people. Other experiments confirm that a shared resource is more quickly depleted by the group if there are some major free-riders than if different players free-ride at different times. In yet another cooperation experiment, a single bad apple lowered the rate of cooperation from about 50 percent to about 20 percent.[17]

The presence of a bad apple incites other group members to stop cooperating (and therefore punish themselves, because everyone gains less with less cooperation). In a social experiment where a lazy, loafing, loudmouthed individual was planted among a group of people who were supposed to cooperate on a task, other team members took on his characteristics.[18] Bad-apple behavior is contagious.

At the core of our aversion to bad apples seems to be an aversion to unfairness. In real-world cooperation dilemmas, unfairness is often evoked, too. But the debate—and one that seems to greatly divide human politics—is over the interpretation of fairness. Should resources be distributed evenly among all group members or proportionally based on some variable such as effort, social status, or access to information? In 1997, the U.S. Senate unanimously passed a resolution (95–0) that said the United States should not sign any international agreement to control greenhouse gas emissions unless it also mandated commitments "to limit or reduce greenhouse gas emissions of Developing Country Parties within the same compliance period." So although President Clinton signed the Kyoto Protocol in 1998, it was never ratified, and in 2001 President George W. Bush made a statement against the protocol because "it exempt[ed] 80 percent of

the world from compliance" and was "unfair and ineffective."[19] What President Bush failed to mention was that
the other 80 percent of the world doesn't produce 80 percent of the greenhouse gas emissions. By contrast, other
countries argue that it is unfair that they should have
to sacrifice economic growth that the United States and
Europe achieved in part through the use of fossil fuels.
Furthermore, the United States was and remains one of
the biggest per capita emitters. In 2011, Canada seemed to
follow the U.S. lead and officially withdrew from Kyoto,
its environment minister suggesting that Canada would
be a part of an agreement only if it included all major
emitters.[20] Bad-apple behavior can be contagious.

Shaming Bad Apples

When we're all in it together, as we are when it comes to
dilemmas like global climate or endangered species, bad
apples cannot be ignored, because their presence jeopar-
dizes the outcome for everyone else. They also cannot
be dealt with easily using guilt, because unless everyone
involved in the problem feels guilty and engages with
that guilt by abstaining from the behavior, the outcome
remains uncertain. It is certainly worth asking what could
make a bad apple turn itself good, but for many collective
problems like climate change, rhinos, or migratory birds,
there simply is not enough time to wait for guilt to prevail.

Collective dilemmas call for regulation by something
more serious than guilt, which is where harsher forms of
punishment come in. In the most serious collective dilemmas, punishment for bad-apple behavior can be severe. In
warfare, military traitors and deserters can be threatened

with capital punishment, because their defection can compromise an entire military action. But if there is no formal avenue for punishment, the group is left with shame as one of its only means of enforcement. The lack of formal avenues of punishment is why shame remains one of the strongest means of coercion in international politics.

In a laboratory setting that used cooperative games, the threat of social exclusion prevented bad-apple behavior from becoming contagious.[21] In our own experiments, the threat of shame made people more cooperative, and they gave more money to the public pool. At the international level, adverse publicity has been used to coerce bad-apple countries into agreements. The presence of bad apples in highly collective dilemmas combined with the lack of other, formal means of punishment gives us reason to consider shaming, albeit in cautious and sophisticated forms.

Given that shame is so tied to the norm it is trying to enforce, it's also important to try to understand what makes norms become the rule. Some norms are easily adopted, others are difficult to change, some shift quickly, others slowly. "How long?" Martin Luther King Jr. asked in 1965 from Montgomery, Alabama, where he assured the crowd that "the arc of the moral universe is long, but it bends toward justice." Just three years later, in Memphis, Tennessee, King was assassinated. Four decades after that, the first black man was elected president of the United States.

Many thought slavery would never end, women would never work, and a black man would never become the U.S. president. Things never stay the same—and because norms are in a constant state of evolution, shame's work is never done. No instruction manual exists on how to guarantee a norm will become normalized; if it did, the world would be a very different place. Some norms, like cooperation, fairness, and honesty, were some of the earliest foundations for shame. But today we have countless other norms—some fleeting and some more solid. Norms are what give shame its role in society. The next chapter explores some ways in which norms become normal, and shame's potential in those processes.

5

How Norms Become Normal

That's just the way it is. Things'll never be the same.
—2PAC, "Changes" (1998)

When Barack Obama appeared on *The Daily Show,* just weeks before the 2008 U.S. presidential election, host Jon Stewart said, "The polls have you up, but then they keep talking about the Bradley effect, this idea that white voters . . . will tell pollsters they will vote for an African American but they won't actually do it." Stewart's point was an important one, because how people say they will behave often doesn't match up with what they actually do. "Yeah, they've been saying that for a while, but we're still here," Obama said. "So I don't know—I don't think white voters have gotten this memo about the Bradley effect."

It is surprising that just talking about the Bradley effect on national television didn't lead to the Pygmalion effect—which is how sociologists describe what happens when the expectations of others lead to a self-fulfilling prophecy. (Pygmalion was a sculptor in Greek mythology who carved a female out of ivory and made a wish that made it come to life.) Studies show that emphasizing the *expectation* of low voter turnout could lead to fewer people voting.[1]

The past several decades have provided research into human behavior, decision making, and norm formation. Research has shown that people tend to calibrate their actions to what they see or hear is common behavior, in the same way they pick up grammar or accents. When visitors to a national forest read signs asking people not to steal petrified wood because a lot of people had stolen wood in the past, theft actually increased.[2]

Evidence also suggests that simply raising awareness is, on its own, a weak instrument for changing behavior (the same might not be true for changing attitudes), while manipulating the "default" choice has contributed to shifts in the norm, without restricting choice. In *Nudge* (2008), behavioral economist Richard Thaler and legal scholar Cass Sunstein argued for changing defaults across society to better represent social aims. Their prize examples were personal savings and organ donations. A U.S. study examining financial saving behavior found that when employers offered optional enrollment in a 401(k) savings plan, 30 percent of employees participated. If, however, the default was reversed and the firm instead offered optional withdrawal from a 401(k) savings plan, 80 percent of employees kept the 401(k) plan (without losing their freedom to choose—employees could opt out with a five-minute process).[3] If the default for a state organ donation program is opt-in—where people have to choose to become donors—the rate of organ donation is much lower than if the organ donation program is opt-out and people must choose to *not* donate their organs. In an opt-in system, fewer people opt to give. In Germany, only 12 percent of individuals give their consent for organ donation, while in Austria, where the default is opt-out, 99 percent of people do. For certain decisions, a more desirable social outcome and a new norm can be achieved without compromising a citizen's ability to choose.

Shaming is another important tool in changing decisions and norms, as we will see in this chapter. A boycott, for instance, can serve two purposes: it can ostracize a business in the marketplace, and it can also focus negative attention and act as a shaming technique. Consider the Montgomery Bus Boycott. The Montgomery Improvement Association (founded in 1955 by Martin Luther King Jr. and Edgar Nixon) chose Rosa Parks (the group's secretary) as the emblem of the campaign, and on December 1, 1955, she stayed in her middle-of-the-bus seat and was arrested for refusing to move to the back. For the next year, there was a boycott of the bus system (helped along by black taxi drivers, who offered discounted rides) until the U.S. district court ruled that racial segregation on buses was unconstitutional. Around that time, King explained that "nonviolent resistance does not seek to defeat or humiliate the opponent, but to win his friendship and understanding. The nonviolent resister must often express his protest through noncooperation or boycotts, but he realizes that noncooperation and boycotts are not ends themselves; they are merely means to awaken a sense of moral shame in the opponent."[4] Shaming was one tool used in the fight to change the norm of who got to sit where—as well as the much bigger norm of discrimination. This chapter explores how norms are established and how they are enforced, and the way shame can contribute to both.

What Are Norms?

There are social norms, legal norms, cultural norms, and religious norms. There are moral norms, like respect for the elderly, and norms of convention, like using silverware. Circumcision is both a religious norm, practiced by Jews and Muslims, and, in the United States, a cultural norm. Being right-handed is another norm, this one with a genetic basis, and characterizes something like 90 percent of the human population. Eating spaghetti with your hands is abnormal, even ill-advised, but it's not immoral. In establishing and enforcing norms, shame finds its supporting role.

To get a sense of the earliest cultural norms, we can look at groups that continue to live similarly to our Pleistocene ancestors. When anthropologist Christopher Boehm examined fifty-three hunter-gatherer societies that he speculated matched most closely the social structures of the Pleistocene era, murder and stealing were forbidden in all of the societies, failure to share and beating someone were norm violations in forty-three of the societies, and an aversion to bullying existed in thirty-four of them. He also found that every single society was egalitarian—that is, absent any social hierarchy, big-shot behavior, or political coalitions.[5] Put another way, bad subjects for reality television.

Not every new norm builds toward some moral pinnacle. Cultures can be capricious and have created many group-damaging behavioral norms along the way, some of which appear baffling.[6] Human societies had (or have) rituals to flatten babies' heads (which Neanderthals also did), bind feet, cannibalize the dead (occasionally even their own children), and mutilate genitals in a suite of preferably unimagined ways.

While humans share some common norms, many norms vary by culture and are in a constant state of flux, which means that shame is, too. Norms differ from place to place and between groups of people. I once heard a U.S. Marine say that no Marine would ever be caught using an umbrella. The Monica Lewinsky affair ruptured the seams of U.S. politics, but it would not have attracted the same scrutiny in many other countries. Writer Elif Batuman quoted an elderly pensioner in Moscow during the Lewinsky affair, who said, "Your Clinton is a young, healthy, good-looking man! Where's the misfortune? Look at our half-dead Yeltsin . . . If we found out [he] was sleeping with a young girl, we would declare a national holiday."[7]

As norms appear and disappear, shame is forced to recalibrate. When a norm disappears, so does the shame that comes with it. Some of the shame of single-motherhood has lifted as the percentage of women who are married has declined in almost every state since the 1980s. In Mexico, single mothers are the heads of nearly one-quarter of households. Gay marriage is a done deal in many Western countries (but illegal in thirty-eight African countries). In 2013, the group Exodus International, which spent thirty-seven years trying to make gay people straight through prayer and psychotherapy, announced it was shutting down.

The fact that shame is so bound to the norm also means we should not blame shame—the emotion or the act of shaming—for making us uncomfortable if what we actually disagree with is the norm that shame is attempting to enforce. In the case of bride kidnapping, a tradition in various parts of the world, in which a man abducts the woman he wishes to marry, the shame the victim and the family of the victim would suffer if she returned home after what is assumed to be consummation (forced, imagined, or, in the rare event, willfully) is often the impetus that forces her to stay with her kidnapper. It is not shame that is ultimately to blame, however, for this situation—it is the society that accepts bride kidnapping as a practice.

How Are Norms Established?

The question of how norms are established is relatively unexplored. If a norm can easily be detected, we know that people are adept at adapting to this majority behavior. If a common sink is filled with dirty dishes, people are more likely to leave their dishes unwashed.[8] In a littered environment, people are more likely to litter themselves.[9]

That means that in some cases, showing the crowd the existing norm—such as how little other teens drink or how much energy neighboring households use—can lead to less overall drinking or energy use. But what if we aren't happy with the norm? What if we believe most people drink too much or use too much energy? This is not about getting the crowd to calibrate its behavior to an existing norm, but rather about getting everyone to reduce their consumption, which means creating a new norm.

Shame is more powerful than guilt when it comes to establishing new norms. Whereas guilt relies upon an internalized norm, shame can be used strategically before a norm has been internalized, especially in the absence of formal sanctions or during the period before formal rules are instituted. Community-led sanitation programs that started in rural Bangladesh in 2000 and then spread to India, Indonesia, and Africa (as documented in the 2011 edited volume *Shit Matters*) connected the practice of open defecation to shame. Many programs began with members of the community doing a transect walk, sometimes also referred to as "the walk of shame," in which the group counted the number of human feces along the route. In some cases, the excrement was flagged with the name of the offender. In other communities, after the transect walk and education efforts to connect human feces to disease, leaders used flashlights on individuals who defecated in the open after dark. The program also attempted to co-opt the existing disgust for using private latrines and redirect it toward open defecation. (Women have usually been the first to be convinced.)

Changing the pathways for disgust can also be a major part of new norm formation, since we now know how important emotions are to behavior. Understanding the meta-norms of a culture, which can be used to anchor and promote new norms, is also important to changing norms. In Western cultures, meta-norms like harm and fairness, also referred to as moral foundations and discussed in detail in psychologist Jonathan Haidt's *The Righteous Mind* (2012), drive a lot of moral behavior, and norms that frame themselves in these contexts probably have a better chance of mobilizing the crowd. Another important component of norm establishment is leadership.

Norm Entrepreneurs

While single individuals do not control social norms, some exert more influence than others. Legal scholar Cass Sunstein calls people with disproportionate power in changing behavior "norm entrepreneurs"—a category that includes Gandhi, King, Rachel Carson, British chef Jamie Oliver, and Thai politician Mechai Viravaidya, a.k.a. Mr. Condom, who popularized family planning in his country. Research on recycling behavior showed that city blocks with a leader who informed neighbors about recycling pickup days recycled more than double the blocks without leaders. Norm entrepreneurs also are capable of using shame effectively because they have the trust and attention of the crowd.

Norm entrepreneurs need not be famous, but they should be respected. Both adults and children prefer to listen, watch, and learn from people with status earned from greater skills and success, expressions of confidence, and experience.[10] Prestigious people also more strongly influence beliefs. In an experiment to test this, two groups of students were given a study that estimated the number of students who cheat. One group was told that a professor had conducted the study and the other group that a student had, although in both cases the estimate itself was the same. Each student then estimated the percentage of their friends who cheated, and students' estimates were significantly higher and more closely matched to the estimates they had read if they had been in the professor condition, showing at least in this case that students were more influenced by professors than fellow students.[11] Prestigious people not only disproportionately affect beliefs through their higher social status but also because they often have a broader social reach. A Chinese researcher pointed out that because Confucius spread the idea that intellectuals should become loyal administrators, it is more difficult to recruit Chinese people into scientific research.[12]

But even with a prestigious or powerful individual,

there are no guarantees about how long it will take a norm to change, or if it ever will. Darwin wrote about a chief who tried and failed to change his tribe's painful habit of knocking out their two upper incisors. Many individuals considered norm entrepreneurs today, like Rosa Parks and Nelson Mandela, were at first considered delinquents. They endured enormous amounts of group scrutiny in their lifetimes, and it took decades for their work to materialize. The three servicemen who helped expose the atrocities of the My Lai massacre in Vietnam were initially ostracized by the U.S. military, which three decades later awarded them medals and invited them to speak to soldiers about ethics.

Norm entrepreneurs also don't have to be individuals. Governments also lead in norm establishment and enforcement, as in the case of the separation of church and state, the one-child policy of Mao Zedong, or the $350 fine for honking in Manhattan. Law professor Eric Posner noted that "official pronouncements play an important role, because officials enjoy the attention of the nation and thus can cheaply create focal points."[13] Scandinavian countries might be weak military powers, but they have been norm entrepreneurs in environmental politics, conflict resolution (particularly during the Cold War), and foreign aid policy.[14]

Norm entrepreneurs can be religious groups, environmental groups, and civil rights groups. They can even be banks. Until the mid-1970s, there was a strong norm to prevent Wall Street banks from financing or advising hostile takeovers—when one company acquires another without the acquired company's approval. The norm against helping hostile takeovers ended in 1974, after Morgan Stanley helped finance the hostile takeover of ESB (formerly

Electric Storage Battery) by International Nickel. At the time, Morgan Stanley was regarded, according to a 1981 article in the *New York Times,* as "Wall Street's foremost investment banker," and its prestige and decision to sponsor International Nickel "was a landmark event that made hostile takeovers more acceptable."[15]

Shame Can Scale

The Wall Street norm against hostile takeovers also shows the power of shame at the group level. Even if institutions cannot feel shame the way an individual can, institutions do often change their behavior to avoid or compensate for the threat of exposure or negative publicity. Before Morgan Stanley legitimized hostile takeovers, according to corporate law professor David Skeel, "a bank that violated the norm would be shamed by its peers" (unless, apparently, that bank was Morgan Stanley).[16] To take a more recent example, the UN Security Council and other groups have shamed a number of companies, based mostly in Europe, for importing minerals from Africa from groups that were directly funding and perpetuating conflict. While shame can be used by groups and against groups to change norms, guilt cannot be used successfully in a similar way.

Shaming can even be used by small, trusted groups against large institutions or even nations, and this is perhaps one benefit of globalization—a tighter global community with participants that are more sensitive to reputation. Nonprofit groups successfully used shaming to convince the U.S. government to stop executing juvenile offenders. As part of their strategy, institutions like Amnesty International pointed out that only seven countries apart from the United States—Bangladesh, Iran, Iraq, Nigeria, Pakistan, Saudi Arabia, and Yemen—had executed juveniles since 1990, and this was not particularly admirable company to keep. In 2005, the U.S. Supreme Court, by a vote of 5–4, outlawed juvenile executions. Shame can scale to the group level and can work quickly when the time is right.

Money and Norms

Another popular research area is the effect of markets on norms, although more is known about market effects on individual behavior than on group behavior. In general, markets have cultivated trust among humans who do not know one another and probably will not meet again. Studies also show that people in more market-integrated economies are more willing to pay to punish unfair behavior.[17] Occasionally, monetary rewards have worked to promote new norms, such as household energy savings, but when the payments stop, so does the energy saving. Taxes can also be effective in changing behavior, as they have been in reducing smoking in many parts of the world and traffic, at least in places like Stockholm. We know that incentives matter, but, depending on the type of incentive, they can matter in unexpected ways.

Introducing markets for specific behavior can sometimes undermine other motivations, which can undermine other human values. One popular example is a study of ten day-care centers in Haifa, Israel. Economists watched parents for four weeks and noticed an average of eight late pickups per week per center. In the fifth week, six of the centers introduced a fine for late pickups—$3 per child for parents arriving more than ten minutes late. The fee would be added to the parents' monthly bill (around $380). In centers that charged a late fee, the number of late pickups immediately increased to twenty per week, more than double the late pick-ups without the fine. Even after experimenters removed the fine eight weeks later and tried to return to the old way, the number of late-coming parents remained high.[18] The market for being late shifted the social norm: the financial penalty was less burdensome than guilt or shame had been. Putting a price tag on behavior most people see as negative can sometimes exacerbate, not temper, a trend.

Further evidence was provided by a 2013 experiment in which participants could decide between saving the life of a real mouse or earning real money. In a nonmarket interaction, 46 percent of subjects were willing to kill a mouse for ten euros. Then experimenters introduced a market with buyers and sellers, in which sellers were given control over a mouse's life and buyers were allowed to offer a price at which to buy the mouse. If the two players agreed on a price, sellers received that price, buyers received twenty euros minus the price agreed upon, and the mouse was killed. (If they did not agree on a trade, earnings for both players were zero and the mouse was saved.) In this market condition, the number jumped to 72 percent of subjects willing to kill a mouse for ten euros or less.[19]

What about using the market to encourage good behavior rather than discourage bad behavior? Harvard University economist Roland Fryer set out to test this idea with a gigantic experiment to see whether paying underprivileged minorities in 203 schools across three cities could cultivate a new norm to study harder in school. In

Dallas, his team paid second graders $2 for each book they read. In New York City, they paid fourth- and seventh-grade students for their grades on ten tests. In Chicago, they paid ninth graders every five weeks for grades in five of their courses. The experiments paid out $9.4 million to around 27,000 students, and each lasted the entire academic year. Before they were finished, Fryer discussed the experiments publicly, including, in 2008, on *The Colbert Report,* where Stephen Colbert quipped, "I love this. This is making the free market make kids learn."

Colbert's supposed love was short-lived. In 2011, Fryer published the results of his work, which showed that financial incentives had little or no effect on academic performance—not for the reading, tests, or courses—and no effect on the students' self-reported effort.[20] Similar results have been reported in programs that pay health care practitioners for healthier patients and programs that pay teenagers not to become pregnant again.

While some norms just are not for sale, we have nonetheless entered a period in which U.S. prisoners can pay for a prison-cell upgrade, tourists can pay to shoot endangered species, and Indian women can be hired as surrogate mothers. Political philosopher Michael Sandel argues that "market reasoning empties public life of moral argument" and that we are now living in a market society rather than a market economy. "The difference is this: A market economy is a tool—a valuable and effective tool—for organizing productive activity. A market society is a way of life in which market values seep into every aspect of human endeavor."[21] Norms have now come up against their market value, when in fact some might (or should) be priceless.

Norm Enforcement

Less research exists on how new norms are established. Instead, there is much more work to explain how existing norms are enforced. Just being watched, for instance, can enforce the norm, and is referred to as the audience effect. Many animals are acutely aware of their social surroundings, and an audience can trigger changes in behavior. It can even happen between species. Experiments also show that dogs steal significantly less food when humans are watching them.[22]

Being watched can also increase the chances of beating tuberculosis—a disease whose symptoms usually disappear within the first two to four weeks of treatment. When patients no longer feel sick, they often stop taking their medication, which can lead to drug-resistant strands of tuberculosis and puts the next person to get infected at risk. To solve this problem, a very simple program was developed, and the World Health Organization has evidence it works: someone must watch the patient swallowing his medications. (An alarm hasn't worked as well as a human.)[23]

Experiments also show that people use less energy at home when they receive feedback, they give more to charity when they think someone will match their donation, and they are more generous after being primed with religious words.[24] Other experiments reduced cheating by three-year-olds (who were told not to open a "forbidden box") when they were alone by telling them that an invisible princess was also in the room.[25] Primed with a pair of eyes, people left behind less waste in a university cafeteria,[26] cleaned more garbage from bus stops,[27] and gave more money to a recipient in a dictator game.[28]

But long-term studies show that the eyes effect weakens over time. An experiment at Baptist churches in the Netherlands showed that replacing closed offering bags with open baskets, which allowed churchgoers to observe one another's contributions as well as the money already donated, increased offerings by 10 percent. Churchgoers also gave fewer small coins. But there was another result: the increased donations elicited by the open baskets petered out as the weeks passed. By week 29, there was no longer an obvious effect.[29] Studies of closed-circuit television (CCTV), which has been introduced in many urban areas to deter crime, show that about half of the CCTV projects studied reduced crime, but not forever. In several cases, such as with the CCTV units in the London Underground, effectiveness was reduced or eliminated in just under a year.[30]

Part of the success of the audience effect is that it serves as a warning that punishment could occur. If that punishment never comes, perhaps the audience effect loses its potency. At all levels of human social organization, punishment is used to enforce norms and is, in circular fashion, a norm itself. Studies show that humans find satisfaction in the punishment of norm violators, and, similar to cash prizes, punishment activates reward-related regions of the brain.[31] The failure to punish can in itself be a shameful thing. Journalist Glenn Greenwald described "the lack of even a single arrest or prosecution of any senior Wall Street banker for the systemic fraud that precipitated the 2008 financial crisis" as "one of the greatest and most shameful failings of the Obama administration."[32] (Greenwald wrote that in 2013; in 2014 one bank executive was actually sentenced to thirty months in jail.)[33]

Punishment plays an important role in establishing

and enforcing norms in nonhumans, too. Experimenters established a norm with rats—not to eat their fifth piece of food—by punishing them with a startling hand clap and vocalization. But if the experimenter left the room, the rats would pause after the fourth piece, stand on their hind legs and sniff the air, and then consume the rest of the food, demonstrating that the norm to not eat the fifth piece was established and enforced through a threat of punishment rather than internalized in the rat.[34]

Punishment is also important in enforcing the norm of honesty. Some systems of communication are more prone to deception than others. Evolutionary scientists Michael Lachmann and Carl Bergstrom have shown mathematically that language is the communication system most prone to deception, because words can be combined to create new meanings, which means there is an infinitely large number of possible lies.[35] As a result, human societies require strong norms for honesty and strong punishment to enforce them, but an experiment with birds, published in 1977 in the journal *Behavior*, was the first to show a clear use of punishment for deception in nonhumans.

Sievert Rohwer, a behavioral ecologist, studied how the norm for honesty was enforced in Harris's sparrows, little high-pitched songbirds that migrate up and down the middle of North America. Any scientist who writes in the methods section of a scientific paper that he is "seriously indebted to the ladies of Crums Beauty School" is someone I want to meet. To find out more about his seminal work using color manipulation, I visited Rohwer at his University of Washington office, which was filled with twelve filing cabinets—marked with names like ornithology, molt reprints, data, and humans.

First, a word about Harris's sparrows. They seem sweet enough, but they actually live under a strict authoritarian regime, with the oldest male birds—referred to as studlies—occupying a dominant position with VIP status. The bigger and blacker the throat patch, the studlier the male, and the higher the priority for accessing food, which is particularly scarce during the winter season. In the spring, when food is abundant, the male sparrows all show the same high-ranking signal: a large patch of black throat feathers. It isn't until the fall, when the sparrows molt and head south, that they replace their throat feathers with a patch that advertises their position in the hierarchy—large throat patches for the studlies, which are often older, and smaller patches for the subordinates, who still spend time with the studlies but only get to peck at their leftovers. Studlies also get priority access to territory, like snowmelt puddles that are close to cover, which is the Saint-Tropez of winter sparrow real estate. On the other hand, Rohwer explained, "subordinates don't have to fight, and they don't have to lose."

But it's easy to produce black feathers. Everyone does it in the spring. So why be honest about your low status in the fall? Dark feathers are cheap, and testosterone is cheap. Why not produce more of both? Rohwer wanted to know why these signals are honest, so his plan was to deceive the flock. "I thought, I'm going to create cheaters," he said. "But I didn't."

For the emasculating, riches-to-rags treatment in which Rohwer needed to downgrade the sparrows in status by shrinking their black throat patches, "peroxide alone did nothing" and "household Clorox completely consumed feather vanes." This was why Rohwer needed the ladies from the beauty school, who he said were "open

to helping because other researchers had asked them how to dye cattle." They managed to bleach the glossy black feathers to a reddish or straw yellow color. The bleached birds became highly aggressive, because other birds no longer recognized them as king of the roost, and attacked other birds three times more than they had before they were bleached. The newly bleached birds weren't getting the recognition they deserved and had to, as Rohwer put it in his paper, live "in a disrespectful world."

But the real drama was to come in what Rohwer thought would be the rags-to-riches treatment. Rohwer added color (using Redken Custom Creme Colour, 151, blue-black) to the throats and crowns of eight winter birds so that they resembled "the studliest of adult males." But sparrow society is a strict caste system, and moving up in rank isn't that easy. With only one exception, the studliest sparrows socially persecuted and excluded dyed birds. Harris's sparrows are usually not seen alone, and none of them came alone to the food patch in Kansas before being dyed. But after the Redken treatment, "four of the dyed birds began coming mostly alone or were relegated to the periphery of the food patch outside of the main flock of birds." Rohwer had created a class of untouchables.

What Rohwer showed was that signals are tested, and there are strict rules about honesty when it comes to moving up in rank. Deceit in sparrows was socially controlled by physical violence and ostracism. (Sparrows don't have a system of exposure that we would classify as shaming.) Also, notice the asymmetry: birds were punished for attempting (not willfully, in this case) to deceitfully move up in rank, but there was no punishment for moving down. What would be the incentive to decrease in rank?

The experiment was clever, but it was also, in some

sense, accidentally cruel. I asked Rohwer if he ever felt bad about interfering with the social order of the Harris's sparrows. "I shot birds for the museum and had no trouble with it. But I also didn't know anything about the individual birds. After I color-banded these birds, got to know them, then manipulated them and watched them suffer at the hand of their own flock, it was devastating," he said. Rohwer's results are widely admired and cited.[36] The study contributed a great deal to our understanding of how social animals promote honesty and, as far as we know, caused no long-term upset to sparrow society.

Given that punishment plays an essential part in norm enforcement, and that punishment can be costly—for both the punisher and the punished—it would be useful for societies to figure out ways to make it cheaper. For sparrows, ostracism was probably less costly than physical punishment. In human societies, shaming can be cheaper than physical punishment, and guilt is even cheaper than shaming, because an audience is not needed to enforce the norm. But, as we saw in this chapter, shame can be more useful than guilt, because of its unique capacity to establish new norms and because it can be more powerful at enforcing norms. Because shame scales, it can also be used on groups and therefore change behavior more quickly. The next chapter examines characteristics that might make shaming more or less effective.

6

The 7 Habits of Highly Effective Shaming

Poor shame, unfairly forgotten, unsexy, and dowdy,
and utterly in need of a makeover.

—WILLIAM IAN MILLER, *Humiliation* (1993)

Consider two case studies in shaming. In the first, a man from Santa Monica, California, publicly exposed a Thai restaurant on Gripe—an app taglined as "a much *better*, Better Business Bureau for the Twitter-age"—for delivering his Thai food thirty minutes late. Gripe, which taps into social networks and review sites, estimated that the complaint hypothetically influenced 260,000 people through the man's nearly two thousand friends and followers. The second case is a website set up by the Political Economy Research Institute, at the University of Massachusetts, Amherst, which used government data to expose the one hundred worst U.S. corporations with respect to air pollution (not only in terms of quantity, but also toxicity, as well as the number of nearby people likely to be affected). The top air polluters of 2013 were Precision Castparts, DuPont, the Bayer Group, Dow Chemical, and ExxonMobil. Large privately held firms were also on the list, like Koch Industries, which ranked fifteenth. As we will find in this chapter, the online list of the hundred most toxic sources of air pollution gets much closer to fulfilling the seven habits of highly effective shaming than the complaint about the delayed Thai food delivery.

Again, shaming is not transparency, which would make everyone's behavior available to the group. Shaming involves exposing only a fraction of the population. What do we mean, specifically, by effective shaming? Shame's performance is optimized when people reform their behavior in response to its threat and remain part of the group, although sometimes, as we saw in the last chapter, shaming is also used to set an example and establish a norm, even when there is no hope of changing the transgressor's behavior, because it lets others know that such behavior will be punished. Even when shaming doesn't elicit the emotion of shame, it can still be effective, as we have seen with group responses to negative exposure. Ideally, shaming creates some friction but ultimately heals without leaving a scar. But part of our discomfort with shaming is that it is difficult to guarantee success, and shaming (like other punishments) can sometimes backfire.

Effective shaming is not necessarily the same thing as permissible shaming. Shaming, like any tool, is on its own amoral and can be used to any end, good or evil. There are also cases in which a certain form of shaming might work to change behavior, but society would find that form of shaming unacceptable. This chapter is not about how to make shaming morally correct. This chapter is about identifying what helps make shaming effective at establishing and enforcing norms, as well as some characteristics that might make shaming more acceptable to the crowd (which is key to effectiveness, given that the crowd is asked to be part of the punishment; for more discussion, see the last chapter). For this purpose we must take an agnostic perspective on individual cases of shaming and consider shaming's general function and form.

Shaming is a potent tool, but its power, as with antibiotics, depends a lot on whether the proper dose is used at the right time. Given that we can all use shame, we should all be interested in using shame selectively and well. The recipe for effective shaming begins with an obvious transgression against a norm, an obvious transgressor, and a desired and achievable outcome. I'll now lay out seven additional habits of effective shaming. I use the term "habits" because these are not rules, but generalities. In brief, here are the seven habits: The transgression should (1) concern the audience, (2) deviate widely from desired behavior, and (3) not be expected to be formally punished. The transgressor should (4) be part of the group doing the shaming. And the shaming should (5) come from a respected source, (6) be directed where possible benefits are highest, and (7) be implemented conscientiously. If shaming sticks closely to these habits, it is likely to play an effective role in changing behavior.

Habit 1: The audience responsible for the shaming should be concerned with the transgression.

Self-evident as it might seem, the audience, which is crucial to effective shaming, should be concerned with the transgression. By this premise, transgressions for norms that are more social in nature should be more likely to attract shaming. In particular, bad-apple behavior, which threatens to destabilize the outcome for the entire group, is a likely target of shaming, because the audience is a direct victim of the behavior. The biggest air polluters put us all at risk of asthma, respiratory infections, heart disease, and lung cancer, unlike the tardy Thai noodles, which afflicted only one customer. The software that alerts your friends and followers every time you hit the snooze button is a less effective use of shaming, because the audience is not bothered by your oversleeping.

That the audience for shaming should also be the victim is why we chose a cooperative dilemma for our shame experiments—the choices of each player in the public goods experiment affected the group. When President Obama told the nation's mayors in 2009 that he would "call them out" if they wasted stimulus money, he couched the warning in similar cooperative terms: that the American people "expect[ed] to see the money that they've earned—they've worked so hard to earn—spent in its intended purposes without waste, without inefficiency, without fraud." The American people were the potential casualties of wasteful spending, so Obama could count on them to be an engaged audience.

The fact that the audience is also the victim has been part of shame's effectiveness in anti-smoking efforts. Health researchers argue that dirty looks and other forms of shaming have been key to reducing smoking rates— just as successful, in some cases, as taxes on tobacco.[1] The negative effects of secondhand smoke continue to mount: heart disease, respiratory infections, asthma, lung cancer, and even hearing loss. In 2004, an estimated 603,000 people worldwide died from secondhand smoke (28 percent were children). Focusing on the secondhand effects of smoking was important to shame's effectiveness, because it moved smoking from a risky but individual choice to a decision that endangered other people.

Contrast smoking with obesity in terms of the effect on the audience. Daniel Callahan, a bioethicist at the Hastings Center, lamented in early 2013 that there is not a stigma like the one against smoking for obesity and called for shaming obese people the way some shame smokers. His article "Obesity: Chasing an Elusive Epidemic" (its title was typeset in a bloated font) got the attention of the media. The press asked whether Callahan was right in arguing that we should shame overweight people into losing weight. But among the many things Callahan got wrong (he implicated "can openers, food blenders and mixers" in the causes of obesity) was the error in equating obesity and smoking as similar dilemmas when obesity lacks the immediate secondhand effects. It is clear that smoking harms bystanders in a more obvious way than obesity, which makes smoking more of a bad-apple behavior. Both have long-term health effects that the entire group arguably shares—but only smoking has immediate negative secondhand impacts that make the audience needed for shaming immediately concerned with the behavior.

That is not to say that the audience must always be an injured or at-risk party, or that the distinction between a behavior that does and does not affect the audience is easy to assess. It is also possible for the audience to feel indirectly harmed, which is part of the reason shaming is also used to stop whaling, land mines, recruiting of child soldiers, and deforestation. Europe's strong reaction to the abuses at Abu Ghraib and the subsequent shaming of American military practices happened not because European citizens were direct victims of the abuse. When transgressions enter a serious moral domain, they are also likely to attract an audience's concern.

Habit 2: There should be a big gap between desired and actual behavior.

Shaming also works best when there is a big gap between attitudes and behavior. As this margin closes, shaming becomes less effective. Many Americans were upset after corporate executives from AIG—which had received $85 billion in taxpayer bailout money in 2008—went on a lavish California retreat, charging half a million dollars to the company expense account. Not only did the public feel they were victims of the transgression (it was taxpayer money AIG was spending, after all), but it was clear that the $150,000 for meals and $23,000 in spa charges deviated from anything considered remotely acceptable—the gap between desired and actual behavior, particularly after a bailout, was large.

The large gap between what we think should happen and what does happen is what makes shaming so effective when it comes to voting. The vast majority of Americans believe people ought to vote. (In a survey of more than two thousand Americans, 89 percent of respondents mostly or completely agreed that it's a citizen's duty to always vote.) Yet only a minority of eligible U.S. citizens actually vote routinely. (Presidential elections are something of an exception, with 62 percent voting in 2008 and 58 percent in 2012.)

Recall the experiment showing that letters that

prompted guilt for not voting led to a 4 to 6 percent increase in voter turnout. A study along similar lines assigned 180,000 registered Michigan voters to different groups that each received a different piece of mail. One letter showed voters their voting records, along with their neighbors' records, with the promise that they would receive an updated version after the election. This letter had the greatest effect on voter turnout: a whopping 8 percent increase. (Normally, no single piece of mail generates anything more than a 1 percent rise in voter turnout.) Social exposure was the ticket to the highest voter turnout[2]—and it worked better than guilt.

The next question is whether the exposure worked better on people who had or had not voted in the past. In another voting study, people in the towns of Holland, Michigan, and Monticello, Iowa, were told via postcard that the names of people who voted would be published in the local newspaper, while in Ely, Iowa, people were told that the names of people who did *not* vote would be published. It was found that the threats of both honor and shame increased voter turnout, but that honor tended to mobilize "high-propensity voters"—those who had voted in at least two of the three prior elections. The shame condition, however, equally motivated both high- and low-propensity voters, defined as those who had voted in one or none of the three previous elections.[3] (It should be noted that "after receiving several calls from local elections officials expressing concerns about publishing names in the newspapers"—another demonstration that effective does not mean acceptable—the researcher decided not to publish the names; however, this decision was made after the election.) Shaming works better when there is a big gap between desired and actual behavior.

Habit 3: Formal punishment should be missing.

Effective shaming should focus on a transgression for which there is no threat of a more severe or formal form of punishment. If there is a formal system of sanctioning in place—if, for example, there were a U.S. law that made voting compulsory (as there is in Australia)—there would be no need to mobilize the audience, and using shame might even be a waste of attention. The crowd itself might even find this to be a waste of its time. A Texas man filmed his neighbor vandalizing his car and not only used it against him in court, but also posted the footage online. Why publicize this crime and further shame the neighbor? The Texas man didn't need the help of the crowd for punishment—there was a system in place to punish the transgression.

When there is no legal mechanism or no enforcement of laws, then shaming moves up the list of options for social control. There were no laws regulating how AIG had to spend the government bailout money, just as there were no laws that prohibited banks from using bailout money for banker bonuses. Financial executives received almost $20 billion in bonuses in 2008, after a $245 billion government bailout. Citigroup proposed to buy a $50 million corporate jet in early 2009, shortly after receiving $45 billion in taxpayer funds. The courts could not find anything illegal in this behavior, but days later, President Obama said that Citigroup "should know better" and called the bonuses "shameful."

Occupy Wall Street was a signal that even though the legal system had not found that the banks had acted outside the law, many people felt the banks had done something wrong. Had more than one lone investment banker gone to prison after the 2008 financial crisis—as hundreds did during the last fraud-filled financial crisis, in the 1980s, which was one-seventieth the size of the one in 2008—we would have probably seen less shaming used against financial institutions and their employees. The Occupy Wall Street movement claimed some modest accomplishments—banks abandoned plans for debit-card fees, the government established a new agency to protect consumers from the financial sector, and the meme of "the one percent" was born. It also served as a reminder that when there are no formal avenues to punishment, society still has shame up its sleeve.

Law professor Toni Massaro has argued against shaming on the grounds that "shaming may convey the message that drunk drivers, child molesters, and the other offenders subjected to these penalties are less than human" and that they "deserve our contempt." In those specific cases, Massaro could be correct. But when shaming is used against bankers or nonvoters or against companies like Google, Amazon, and Starbucks for their offshore tax havens, it's not because society sees them as "less than human" (although in the case of the corporations, they might be) but because there is no alternative. For the same reason, states are left with shaming delinquent taxpayers because, unlike the federal government, states have almost no other means of recourse. (The State of California can confiscate only second homes and luxury vehicles, and only after a big legal brouhaha.)

It's not that shaming is preferable; it's just that, in some cases, shaming is all we have. When fishermen in southern Chile are seen in areas that the community has designated as marine protected areas, other fishermen write the transgressors' names on a big sign in town that says

LOS CASTIGADOS ("THE PUNISHED"). The fishermen are group-policing the area and using shame because there are no formal sanctions. International law instruments like the Universal Declaration of Human Rights have no formal avenues for punishment, so shaming is one of the main tactics of enforcement. Kenneth Roth, executive director of Human Rights Watch, wrote, "The strength of organizations like Human Rights Watch is not their rhetorical voice but their shaming methodology—their ability to investigate misconduct and expose it to public opprobrium."[4]

Sometimes formal punishment will never be implemented, and shaming can encourage better behavior. It is not likely, for instance, that highly salty or otherwise unhealthy foods will ever be outlawed—nor that we would want them to be. Yet we also grapple with the problems associated with unhealthy diets. Research shows that shaming might help reduce consumption of unhealthy foods—not by shaming the consumers, but by shaming the foods, that is, singling out the most egregious products that they might purchase. Researchers labeled some foods at a hospital kiosk "less healthy" and sales of healthier items increased by 6 percent.[5] Beginning in 1993, the Finnish government required a label on foods with high salt content, which led to a significant decrease in overall salt consumption.[6]

Habit 4: The transgressor should be sensitive to the source of the shaming.

The threat of shame works best when the transgressor is sensitive to the group doing the shaming. Moral philosopher Bernard Williams wrote in *Shame and Necessity*

(1993), "The basic experience connected with shame is that of being seen, inappropriately, by the wrong people, in the wrong condition." I would change that sentence to read: "by the *right* people, in the wrong condition." Shaming campaigns repeatedly make the mistake of singling out people or groups that do not consider themselves part of the broader crowd or broader norms that the shaming campaign represents.

The sensitivity to the source of shaming is, according to law professor Saira Mohamed, why President Clinton took political action in Haiti but not in Rwanda or Darfur. After the first democratically elected president in Haiti was ousted in 1991, violence in the country escalated. Clinton responded by stating that no Haitians would be allowed in American territory. But a shaming campaign ensued, led by some of Clinton's most loyal supporters (it included a hunger strike by TransAfrica Forum director Randall Robinson and a full-page advertisement in the *New York Times*), and as a result Clinton changed his strategy, stepped up sanctions, and (with the support of the UN Security Council) deployed U.S. troops to Haiti.[7] Clinton would not offer the same support for other, even more violent conflicts to come in other parts of the world.

Consider also the shaming of nonvoters in the *Tennessee Tribune,* which, just before Election Day in the 2006 and 2008 federal elections, published twenty-eight pages of the names and addresses of people from predominantly African American neighborhoods around Nashville who had not voted. Afterward, Stephen Dubner wrote on his *Freakonomics* blog that "it was a black newspaper . . . that did the outing, and most of the non-voters they outed were black. You can imagine the uproar if the paper wasn't a black paper and the voters were. But it seems it's always more acceptable for a group to criticize itself from within than to be criticized from without."[8]

Greenpeace's Carting Away the Oceans campaign further illuminates why it is important to consider in- and out-groups. Since 2008, the nonprofit Greenpeace has ranked twenty of the largest U.S. supermarket chains according to the sustainability of their seafood. For several years, the retailer Trader Joe's ranked very low and seemed unresponsive to the Greenpeace reports and newspaper headlines. In 2009, Greenpeace volunteers coordinated telephone calls or demonstrations (and sometimes both) at Trader Joe's stores across the nation. The customers pressured the managers of the individual stores, who finally convinced Trader Joe's CEO that their seafood needed to change. The chain's sustainability score has now increased from a 2 (out of 10) in 2008 to a score of 7 in 2013—and Trader Joe's jumped from fifteenth to third place on the list of retailers. Several other large stores, such as Winn-Dixie, have also repeatedly ranked low on the list, yet Greenpeace didn't single out those stores as they did Trader Joe's, partly because their customers are not as obviously aligned with Greenpeace as Trader Joe's customers are.

Greenpeace went after Trader Joe's because it understood that the retailer would be more sensitive to the shaming. "Winn-Dixie is definitely not off the hook," said John Hocevar, Greenpeace USA's Oceans Campaign director. "We've been able to put a decent amount of pressure on them with the scorecards and embarrass them at meetings, but they are a little more difficult than some, because they are predominantly based in the Southeast. It's easier for them to imagine that they can withstand the pressure because they're not in the center of our stronghold of support. We can still fight them; it will just take more resources."

To keep shaming sustainable into the future, it is also important that individuals or companies who are exposed are given a chance to reintegrate with the group. Some of the most effective shaming can be followed with honor to reward changes in behavior. The Greenpeace seafood campaign also honors the stores at the top of the list, as well as the ones that move up in rank. "There are a lot of organizations working with supermarkets, and they deserve some of the credit. But few other organizations have much of a stick," Hocevar explained. "And you can only get so far with carrots alone. We have both. We can offer people the opportunity to say, 'We scored number one in Greenpeace's report,' and at this point three different supermarkets have bragged about achieving that."

Honor was also used alongside shaming during National Public Radio's pledge drive in 2011, when *This American Life* host Ira Glass asked listeners to turn in their "friends, family, and loved ones who listen a lot to public radio but have never pledged money." (Public radio is a perfect example of a public good, because no one can be excluded from its benefits.) Glass called up several of the accused and, using only their first names, publicly shamed them for not contributing to the pledge drive and told them who had turned them in. Yes, people were being called out for not giving money to public radio, but they were turned in by their closest friends and were phoned by a famous radio host. A spoonful of honor helps the shaming go down.

Habit 5: The audience should trust the source of the shaming.

For shaming to have traction, it should come from a source that the audience respects. A study that looked at Russian companies between 1992 and 2002 found that shaming a CEO of an underperforming company worked (meaning the CEO resigned or changed company policy) only if the exposure was in an American or British newspaper, like the *Wall Street Journal* or the *Financial Times*. Exposure in Russian newspapers seemed to have no effect, because, according to the study's authors, those papers lack credibility.[9]

There is a chicken-or-the-egg phenomenon, because it is not clear whether one must be trustworthy to punish or if one gains trust by punishing. After anthropologist Polly Wiessner analyzed the conversations of the !Kung hunter-gatherers in Africa, she found that three-quarters of the criticism came from just one-quarter of the "strong" individuals, and that those strong individuals punished twice as often as those judged to be normal or weak.[10]

The Daily Show's Jon Stewart is someone who has gained trust through his selective use of punishment. Instead of shaming (mostly by ridiculing) the easy celebrity figures, the show chooses subjects who, as one producer described, "deserv[e] to be targets." The new focus on government and politics gave rise to Stewart's popularity. In 2008, *New York Times* columnist Michiko Kakutani asked, "Is Jon Stewart the Most Trusted Man in America?"[11] A 2009 *Time* magazine survey found that the majority believed Jon Stewart was indeed America's most trusted newscaster (with 44 percent of the vote, more than any other news anchor).

Prestigious individuals are more trusted to punish, and less prestigious individuals are less trusted to punish. "No man can lead a public career really worth leading, no man can act with rugged independence in serious crises, nor strike at great abuses, nor afford to make powerful and unscrupulous foes, if he is himself vulnerable in his private character," wrote Theodore Roosevelt in his 1913 autobiography. That's why Jason Russell, caught naked on the streets of San Diego (a mental breakdown), damaged the reputation of his Kony 2012 campaign, which aimed to expose young people to militia leader Joseph Kony's abuses in Uganda and to have Kony arrested.

Shamers cripple their efforts when it is revealed that they are partaking in the behavior they are shaming. The United States undermined its leadership on human rights when abuses at Abu Ghraib prison were revealed. Former Colorado pastor Ted Haggard undermined himself, his family, his church, and, some would say, conservative politics when he preached so ardently against gay marriage but then got outed by his call boy of three years. Dan Savage wrote in the *New York Times* in 2006 that "today it is arguably more shameful and damaging to be a hypocritical closet case than it is to be a sex worker."

Punishers should also be wary of punishing too much or unjustly. In public goods experiments, people reward only punishment that is deemed as justified. In !Kung society, when punishment is deemed unfair, the punisher might be told they are acting jealous or out of turn or be told they are an "angry, sharp, or biting thing," not unlike Lewis Carroll's quick-to-decapitate Queen of Hearts. Too much punishing can make a punisher feared, not trusted, and lead to shaming having the opposite of its intended effect.

Habit 6: Shaming should be directed where possible benefits are greatest.

Given the finite amount of attention the audience has to give, and the problem of getting and maintaining that attention, shaming should be used sparingly, so as to maintain its power. Frivolous shaming can be a distraction from other transgressions that mean more, and a misuse of the audience's attention. Effective shaming requires strategic thinking about the transgressions that matter most, the most responsible transgressors, and perhaps which transgressors are most likely to change and which solutions are most likely to be adopted. From 1987 to 2010, CalPERS, the multi-billion-dollar pension fund that manages benefits for California state employees, named and shamed companies that had poor shareholder performance or corporate governance, but it named only firms in which it held at least $2 million worth of stock[12]—firms that, if they changed, would justify the shaming effort.

It is important to understand how responsibility is distributed. In the case of fossil fuels, demand is diffuse, but supply is concentrated—so it might make more sense to expose companies for bad behavior. In the case of shark fin consumption, supply is diffuse (sharks are caught by fishermen all over the world), but demand is concentrated among the Chinese elite, so it might be more sensible to focus on demand. The Rainforest Action Network could not get traction with coal companies doing mountaintop removal in Appalachia, so instead it traced the responsibility to the nine banks that were the primary lenders to the companies bulldozing the tops of mountains in pursuit of coal, and shamed them year after year for being involved (by the fifth annual coal finance report card in 2014, Wells

Fargo and JPMorgan Chase both had committed to end their financial relationship with mountaintop removal coal mining companies). After a factory collapsed in Bangladesh in April 2013 and killed more than eleven hundred people, the focus was less on the brands manufacturing their clothing there and more on the government of Bangladesh, for its poor safety standards. In its work to expose poor prison conditions, Human Rights Watch avoids trying to change infrastructure and instead focuses on improving the way prisoners are treated, because they see this as a more viable outcome.

To maximize effectiveness, it often can be better to focus on institutions, companies, or countries rather than individuals. Instead of shaming individual consumers who buy Chilean sea bass, Greenpeace shames the largest retailers that sell them. Rather than shaming obese people, shame could be directed at the companies whose profits grow with our waistlines, which is how writer Michael Moss directed his attention in *Salt, Sugar, Fat: How the Food Giants Hooked Us* (2013). Small changes by big groups can make a big difference.

But groups are not always the entities with the biggest impact. Individuals in California owe more in back taxes than corporations, and many of the debts of single households exceed those of single businesses. When the impact of households exceeds businesses, it might be appropriate to focus on individuals. The habit of focusing shaming where it matters most can often mean focusing on the most powerful, including politicians. In 1969, American environmentalist (and expert shamer) David Brower started the League of Conservation Voters, which began singling out members of the "Dirty Dozen"—candidates for Congress who had voted against environmental legislation and were running in races in which a shaming campaign had a serious chance of changing the outcome. Brower knew that it was important to focus not only on the worst environmental candidates, but those for whom negative exposure could make a difference in the race.

Habit 7: Scrupulous implementation.

When designing an effective act of shaming, it's important to consider how it will be implemented. What is the exposure-response relationship? What is the optimal frequency of exposure? What style will attract the most attention and is the most acceptable? Another important aspect of shaming is that the threat can be more useful than the act. This is part of the success of the California tax-delinquent shaming policy: people receive letters six months in advance of their name's listing and have the opportunity to avoid the actual shame.

Research shows that in some cases, like voting, a very small dose of shame can lead to changes in behavior for years afterward. A follow-up voting study tracked more than one million registered voters who received shaming mail of the types discussed earlier and found that the letters continued to statistically increase voting one, two, and up to four years later.[13] Other times, shaming requires a plan for the future, and sometimes even maintenance and repetition. Greenpeace understood that a one-off seafood retailer scorecard would have very little effect. As of this writing, Greenpeace has released seven scorecards since the campaign began, in 2008. "We look at least three years out and develop campaigns that we fully realize can take ten years or longer to win," said John Hocevar. "At some point, we get into the area of diminishing returns and it may make more sense to move into a different sector rather than wait for the last retailer to get a passing score." When Greenpeace stops, will the retailers go back to their old ways? "One of the things we thought was important was not just to drop a specific product but to develop policies to guide purchasing decisions in the future and that these are publicly available," Hocevar said. "When information like this is available to their customers, it makes it more difficult to backslide." Customers will ideally help ensure the retailer behavior is maintained to their (and Greenpeace's) standard.

Shaming is far less effective without an audience that engages with and supports its use. This is where shaming becomes a moving target, because what the audience sees as permissible is always changing. Legal philosopher and shame expert Martha Nussbaum wrote about her alcoholic mother in *Hiding from Humanity: Disgust, Shame, and the Law* (2004) and thought about what would have happened if she had been arrested for driving under the influence and required to get a special license plate on her car (a shaming punishment used in several U.S. states). Perhaps the shaming would have been effective in that Nussbaum's mother would have been dissuaded from drunk-driving again (in fact, one Florida judge said that such license plates led to a 33 percent decline in drunk driving incidents), but, according to Nussbaum, that would not matter. "I know that such a penalty would have broken her spirit," Nussbaum wrote. "It would be a cruel state, with deficient respect for human dignity, that would string up someone for public viewing in that way rather than offering treatment for the underlying problem, together with protection for privacy and dignity." In other words, what is "effective" can change if we broaden our definition to include other considerations.

Finally, it's not enough to simply release information. The negative exposure should be seen, understood, and engaged with by an audience that matters, which means further considerations of the implementation. When New York City created a list of the city's worst landlords, Craigslist linked directly to the website from its apartment listings, so that the information on transgressions was available at the time renters might be making a decision. In the case of the late Thai delivery, the Gripe app was well designed and tapped into large networks of people. (We can't be certain that the Thai restaurant improved its service as a result.) The air-polluters website is not as well designed or as good at accessing social media, but that shaming certainly got more media attention. No matter how it is implemented, shaming will not work if there is no audience and if it doesn't get anyone's attention. Shaming is nothing without an audience, and the Internet is now where the largest audience gathers. The billions-strong crowd, combined with the low cost of information sharing, is another way in which shaming today is fundamentally different from what it was in the past, which is why online shaming is the subject of the next chapter.

7

The Scarlet Internet

In any light, man's further task is Jovian. That is to
learn how best to live with these powerful creatures
of his mind, how to give their fecundity a law and
their functions a rhythm, how not to employ them in
error against himself—since he cannot live without
them.

—GARET GARRETT, *Ouroboros; or,*
The Mechanical Extension of Mankind (1926),
quoted by GEORGE DYSON, *Darwin Among the*
Machines (1997)[1]

"I wouldn't suggest just any Dunkin' Donuts," Matt Binder
said after I thanked him for choosing this midtown New
York City location with a second floor where we could
talk. Binder is someone who thinks strategically, whether
it is conscious or not. (Despite his obvious online lifestyle,
Binder never checked his phone during our fifty minutes
together.) A slight, soft-spoken twenty-something born
and raised in Queens, Binder is a modern vigilante. He
is the lone force behind a Tumblr blog dedicated to pub-

lic shaming—which draws primarily from Twitter for shameful behavior. "People are saying these things on Twitter, Facebook, and e-mail, but Twitter is the easiest to search," Binder explained.

At first, Binder used his blog to post funny misspellings, like "I love the smell of my dad's colon," with a series of screenshots of tweets from people who had meant to write "cologne." Then he got interested in politics and hypocrisy, calling out someone who wrote, "If Obama wins this election it's only because all the lazy bums will come out and vote for him so they can stay on welfare #getajob." Twenty-four minutes later, the same person posted, "Someone help me get a job, these looks don't put money in my pocket."

The scale of ignorance, racism, and cruelty on Binder's blog (publicshaming.tumblr.com) shows that his work is less about exposing individuals and more about holding a mirror up to aspects of American culture. His blog picked up traffic after he curated tweets from geographically challenged Americans who thought the two Boston Marathon bombers were from the Czech Republic. (They were from Chechnya.) Since then, he has shamed streams of people who, after the Supreme Court ruling against anti-gay marriage legislation, used the phrase "It's Adam and Eve not Adam and Steve," people who posted racist tweets about the Mexican American ten-year-old who sang the national anthem during the 2013 NBA Finals, and people who thought we should divide up the $550 million Powerball lottery among the 300-plus million Americans and . . . give each American a million dollars. (According to Binder, highlighting bad math skills "lightens the material.")

As early as 1962, philosopher Marshall McLuhan

referred to new electronic media as "a global village,"[2] but it wasn't until the explosion in digital technology that things really picked up. The global village can be diverse and supportive, but it can equally be a small-minded rumor mill or a place where people—as individuals, like Edward Snowden, or as loosely organized collectives, like Anonymous—might take the law into their own hands. The Internet town square isn't a central meeting place, and it isn't physical. No one is tied to a pillory, clapped into the stocks, burned at the stake, or publicly hanged. But there are severe forms of online shaming.

Some online venues, like Binder's Public Shaming blog, are cautious and calculated in their use of shaming. Binder espoused the usual dogmatic belief in a free and open Web where anything goes, which is widely held in the tech industry, but when I pressed him on the subject, he explained his many rules for his own blog. He first tries to ensure that the material is from a sincere Twitter account and that it is not sarcastic. He tries to avoid tweets from trolls. ("If your Twitter handle is 'I hate Obama,' and you're posting something racist, why should I give you more attention?") Binder doesn't remove Twitter handles when he reposts tweets, so people are not anonymous, but because he uses screenshots of the material, his site generally is not searchable (in part because it would require extra time, but Binder is not sure he would want it to be searchable anyway), and this helps protect the individuals and focuses the attention more on the behavior. Others online are less strategic. Given that the Internet is the easiest platform for shaming with the largest possible audience—a potential crowd of 2.7 billion people (and growing)—it's useful to consider what shaming online means for all of us.

The New Frontier

In seventeenth-century Boston, Hawthorne's Hester Prynne ambled through the streets with a scarlet letter "illuminated upon her bosom." Today, the U.S. government might not be interested in her affair (unless she was a U.S. politician), but if Prynne were Chinese, her marriage record might be online. Chinese officials began making marriage records available in 2011, starting with Beijing and Shanghai, with plans for all of China by 2015.[3]

The all-seeing, omnipresent government makes most people uncomfortable, which was true long before the Internet. Writer Milan Kundera, who immigrated to France from a "surveillance-riddled Czechoslovakia," wrote that "when it becomes the custom and the rule to divulge another person's private life, we are entering a time when the highest stake is the survival or the disappearance of the individual."[4] Kundera was referring to state surveillance, which the Internet allows with increasing ease.

Despite what might be implied in its name, we now know that the U.S. National Security Agency collects data on individuals all over the globe, even in virtual worlds like Second Life. Police departments around the United States have experimented with different forms of shaming on different social media platforms for every type of criminal. Governments not only are watching us, but are developing policies and platforms that threaten to expose, and thereby shame, public offenders. But large-scale surveillance and the potential shaming that comes with it are no longer reserved for just governments.

Digital technologies now provide a platform conducive to *anyone* exposing behavior, not just the state. The Hester Prynne of today need not worry about the government, but she might instead find herself the object of contempt on a Facebook page created by her cuckolded husband. He might add her name and photograph to CheaterVille.com (whose bald motto is "Look Who's Getting Caught with Their Pants Down"™). He might post naked photos of her online. He might decide to use social media to shame Prynne's paramour. (In June 2011, a judge in England dismissed two charges of harassment against a London man who had used social-networking sites to shame his wife's lover.) In other words, the Hester Prynne of today isn't necessarily any safer from shaming.

There are nongovernment websites displaying cheaters, mug shots, deadbeat dads, and sex offenders. A neighborhood group in Leicester, England, posts videos of people caught littering, and removes them only if the "litter louts" are identified and pay their fine. A father made his teenage daughter post a video to Facebook in which she admitted to how young she was and apologized for deceiving boys. A colleague told me that after an argument, his teenage son edited his father's Wikipedia profile to say he was a pedophile (ah, New Yorkers). Consumers use their social networks to expose bad services and products, but it cuts both ways: a restaurant in Los Angeles uses Twitter to shame people who do not show up for their reservation. Today, there is a whole reputation-related Internet vocabulary, such as digital footprint (what's on the Web about you), digital dirt (the bad stuff about you online), sock puppet (an online identity used for purposes of deception), dooced (to lose your job for something you said on your website), and doxing (the act of revealing personal information about someone online).

But isn't this just an Internet version of vigilantism,

which, like shaming itself, changes with each new set of communication tools? Is online shaming really so different from being tarred and feathered or being featured in the tabloids? One major difference between shame online and shame past is the speed at which it can happen. Another is that, with the Internet, it's no longer necessarily clear who is doing the shaming. Anonymous, an informal and anonymous collective of online activists and protesters, hacked accounts and leaked a video online of two Ohio high school football players joking about having raped a girl, which, defense attorneys worried, undermined their right to a fair trial. Anonymous issued a statement that "they" shared the worry about a fair trial, too—because of cover-ups by officials and because the boys were star athletes. A judge sentenced both boys to prison time.

The speed at which information can travel, the frequency of anonymous shaming, the size of the audience it can reach, and the permanence of the information separate digital shaming from shaming of the past. In this new global panopticon, we must be mindful of shame's power and its liabilities. It is difficult to imagine anyone ever acquiring pre-Internet levels of privacy, but there are now looming questions about whether there is a right to privacy and, if so, where it begins and ends. It is also true that with regards to these new platforms, many of the old arguments against shaming become less useful. (The term "Internet" is not indexed in Martha Nussbaum's book against state shaming.) Legal scholars who have argued that shaming cannot be effective in highly mobile, anonymous, and urban societies haven't spent enough time online.

Does Shaming Want to Be Free?

It's difficult to keep track of the billions-strong population of today. "We demand of a community that it should be stable, in the sense that it does not conduce to the disasters of war, and at the same time progressive enough to bring about a continual improvement in living conditions, since failure to achieve this results in discontent and revolution," wrote electrical engineer and physicist Hannes Alfvén in his 1966 *The Tale of the Big Computer* (under his "monozygotic relative" and pseudonym, Olof Johannesson). "To build such a society is a very difficult problem; so difficult is it, indeed, that it exceeds the capacity of the human brain, and can be resolved only with the help of computers."

Digital technologies allow us to keep track of one another in new ways. On top of storing and transmitting gargantuan amounts of data, they allow information to move at unprecedented speeds. Unlike the gossip of the past, which was spoken or printed, Internet gossip is fast, far-reaching, set in digital stone, and often searchable. Even if shameful tweets are deleted or disagreements are resolved, there is often a lingering digital trail. With technology like the Wayback Machine, an Internet library that archives digital information, including webpages, nothing on the Internet need ever fully disappear. Employers today can check the Internet to see if their potential hires should be disqualified by any indiscretions made at any point in time, intentionally made public or not.

Shaming has never been so cheap or so accessible. Under these conditions, it's not surprising that a lot of online shaming represents a low-risk form of moral engagement. Would the people who posted the 215,383 negative Twitter messages in a single day drawing attention to the Susan G. Komen decision to stop funding Planned Parenthood have turned out on the streets for an actual protest?

Shaming is also so cheap because it can be anonymous, which allows the online mob to more easily do what law professor Toni Massaro worried about: express "contempt for the offender" rather than "mere opprobrium for the offense."[5] A police department in New Jersey posted the names and photos of people charged with all types of crimes on its Facebook page, but comments on the site quickly became "off color and racist." The police department then opted to post only the names and photos of those arrested on felonies, with the exception of drunk driving, and they shut down the comments section entirely.[6] When the Minnesota Department of Public Safety tweets about its drunk-driving arrests, they do not include the names of the individuals. California law exempted the home addresses of elected or appointed officials from being made public on its "Top 500 Delinquent Taxpayers" page, for fear of retaliation toward those officials from the crowd.

Shaming is also cheap from a financial point of view. In a 2005 analysis in advance of the launch of its tax-delinquent Web page, the State of California estimated a one-time cost of $162,000 for the creation of the webpage and database, with ongoing annual costs of approximately $131,000. The report also speculated that the page would result in collections of approximately $1.6 million—well worth the upfront costs. They were wrong. Since its inception in 2007, the listings have led to the recovery of more than $336 million in tax revenue—more than two hundred times what they expected.[7] In 2006, the Wisconsin state government began posting all names of people and companies owing in excess of $5,000 in taxes. This led to the recovery of more than $108 million in less than five years, more than fourteen times what officials predicted.

Because it's both inexpensive and effective, people are figuring out all sorts of new forms of online shaming. Union picketers now place cameras with signs warning those who cross the picket lines that their photos, names, and addresses will be put online.[8] After an online video of the founder of GoDaddy killing an elephant got negative attention, PETA launched a campaign urging people to close their GoDaddy accounts, and a competing company offered to donate a dollar for every new account opened, raising more than $20,000 for charity. (The founder of GoDaddy said he had only been helping African villagers whose crops are eaten by elephants. He had apparently "helped" them five times by killing five different elephants, and as PETA's president pointed out, there are other, nonlethal ways to help African villagers.)

A common early form of online shaming was "sucks" sites. A sucks site can legally use a company's trademark to criticize the company, as long as it makes it clear to visitors that the site is not sponsored by or affiliated with the actual company. But, as a *Wired* magazine article from 2000, titled "Legal Tips for Your 'Sucks' Site," pointed out, there always remains the possibility of a lawsuit. Rich businesses now have teams of lawyers to fight this sort of masquerading, as do really rich individuals. But most of us are not afforded the same level of protection.

The digital world is susceptible to deception because identities are so easy to fabricate. Because of the high levels of identity theft and fraud on social-networking sites, sites usually offer a way to report it. Facebook has a procedure for reporting bad accounts, which include accounts that "pretend to be you or someone else, use your photos, list a fake name, or don't represent a real person (fake accounts)." While perusing Gripe, I stumbled on a complaint about a tax service in Indonesia, and the profile picture used for the complaint was a stolen photo of my colleague at New York University (NYU).

Despite all the potential liabilities of online shaming, Internet scholars like Clay Shirky and Jonathan Zittrain argue that strict laws against online shaming would be difficult to implement without also restricting fruitful criticism and activism. The line between punitive and activist shaming by the crowd can be difficult to draw. The primary concern in the early days of the Internet was protecting both freedom and privacy without too much government oversight. In 1990, Mitchell Kapor cofounded the Electronic Frontier Foundation to "assure that these freedoms are not compromised." In a September 1991 article in *Scientific American* about online civil liberties,[9] Kapor described some of his primary concerns, such as disproportionate punishment of hackers. "A system in which an exploratory hacker receives more time in jail than a defendant convicted of assault violates our sense of justice," wrote Kapor. Those early concerns would become real problems.

Disproportionate Punishment

One consequence of digital shaming's inexpensiveness is that it is used frequently and viciously. Shaming by the online crowd involves the problem of proportionality: minor offenses caught on film and widely distributed can elicit greater punishment than major offenses that go unrecorded, resulting in an inherent and probably unavoidable unfairness. When the owners of Lola the cat found her in a garbage bin outside their home in Coventry, England, they checked the footage from their closed-circuit video camera (hidden cameras are the new home accessory) and saw that a person had put the cat in the bin. The Lola owners posted the video online and set up a Facebook page titled "Help Find the Woman Who Put My Cat in the Bin."

Within days, someone online identified the woman as "M.B." (I am abbreviating identities in this chapter so as not to add to the opprobrium), who would then receive thousands of online death threats. (The "Death to M.B." Facebook page was deleted due to terms-of-use violations.) More than twenty thousand people "liked" the Facebook page "M.B. Should Be Locked Up for Putting Lola the Cat in a Bin," while the page "Cats Unite Against M.B." was much less popular. Even scarier were the real people who gathered outside M.B.'s home. In court, M.B. pleaded guilty and was fined £250 (about $400) and banned from owning a pet for five years. In justifying her fairly minor punishment, the judge said she had taken into consideration all the negative publicity M.B. had received.

Another case of disproportionality occurred after a video of seventh graders bullying a sixty-eight-year-old school bus monitor in Greece, New York, went viral. Yes, the video was awful. But the teens received death threats, and we can all agree that no thirteen-year-old should die for something he said, on the basic premise there wouldn't be many thirteen-year-olds left. One teenage boy received more than one thousand calls and one thousand threatening text messages. In this case, the disproportionality continued with the lottery-jackpot-like ending for the bus monitor. A man in Toronto used an Internet crowdsourcing platform in hopes of raising $5,000 to send her on vacation—and wound up raising more than $700,000. Again, the online audience is large.

Shamelessness Online

It's not just the shaming that's a problem online, but also the shamelessness. *New York Times* reporter John Markoff was there at the beginning. He published an article in 1993 titled "A Free and Simple Computer Link," in which he described the Web as "an international string of computer data bases that uses an information-retrieval architecture developed in 1989 by Tim Berners-Lee." He also mentioned a partnership to distribute a cheap software package called Internet-in-a-Box. Markoff told me over e-mail, "What I remember is going from a naive belief that the web was a 'utopian abode' to a realization that the online world was full of trolls and other nastiness and that people behaved online in ways they would never behave in face-to-face encounters. I think I learned a lot from Vernor Vinge's *True Names* [the first science fiction novel about cyberspace, published in 1981] but it took awhile to realize the life was imitating art."

There is a surprising amount of behavior online that we would not find in real life, which is formally referred to as the "online disinhibition effect." If you were reading blogs in the mid-2000s, you probably know of the vitriolic nature of online debates between atheists and creationists. It's not just religion that gets people worked up, as almost any strand of comments on YouTube attests. Moderators on food blogs know things will get ugly the minute the topic of children in restaurants surfaces. Yet few of us have seen anyone go bananas on some unrelated rowdy child at a restaurant. All the real-life atheism debates between online personas were relatively calm and civilized.

Anonymity is one reason people are disinhibited. We expect different behavior on sites that require effort in creating and maintaining a reputation than on sites where users can engage with easily obtained and destroyed identities. Jaron Lanier wrote, in *You Are Not a Gadget* (2010), "You could conclude that it isn't exactly anonymity, but transient anonymity, coupled with lack of consequences, that brings out online idiocy. Participants in Second Life [a virtual world] are generally not quite as mean to one another as are people posting comments to Slashdot [a tech news website]. The difference might be that on Second Life the pseudonymous personality itself is highly valuable and requires a lot of work to create."

Instead of government oversight, individual sites (like Binder's) and institutions are making their own rules. Again, Mitchell Kapor from 1991: "It would be unthinkable for the telephone company to monitor our calls routinely or cut off conversations because the subject matter was deemed offensive." But, unlike the telephone, this is no two-way conversation, and one bad apple can spoil it for the whole bunch.

Not many people want to spend time in a place where jokes are made about the latest rape or school shooting. Newspapers like Chicago's *Daily Herald* regularly delete comments from people who harass others or joke about tragedies. Before they ended anonymous comments in 2013, the *Huffington Post* had forty human employees who watched for and deleted posts that were racist, homophobic, or hateful in other ways.[10] The *New York Times* moderates all comments. Some newspapers now require commenters to use a Facebook profile when leaving a remark, in the hope of preventing offensive behavior. In late 2013, YouTube began requiring people to log in to Google Plus to make a comment (Google owns YouTube), in part to cut down on the negative comments (as well as, one assumes, to collect more data). Google lowered the page rankings of sites that post mug shots and then charge money to remove them. It's not the state, the law, or liability driving these decisions, but a concern about the culture and atmosphere each website aims to cultivate. Kapor might have thought it "unthinkable" that the telephone company would monitor or censor our calls, but that's exactly what many online companies are doing.

Sometimes shamelessness meets shaming. In one such case, a man I'll abbreviate as H.M. (not to protect him, but in order not to give him any more attention, which seems to be what he wants) created the site Is Anyone Up? in late 2010. Anybody could anonymously submit nude photos, mostly of ex-girlfriends—this was referred to as "revenge porn"—which H.M. would post. When he received an e-mail from one woman's lawyer, H.M. responded with a photo of his penis. *Rolling Stone* called him "The Most Hated Man on the Internet." The site remained up until spring 2012, when Bullyville.com—a site dedicated to helping people get past bullying—bought it and took it down.

When Digital Gets Physical

Shaming in the digital realm doesn't necessarily stay digital. In August 2011, a woman whose nude photos had been posted without her permission on Is Anyone Up? confronted the site's founder and stabbed him with a pen. The wound apparently required stitches (but not much sympathy). In China, human-flesh search engines, a sort of Chinese equivalent of the Anonymous collective, induce people to do online research in order to seek out and expose transgressors in real life. The search also often encourages the system to enact formal punishments, demanding, for example, that the transgressor be dismissed from his or her job. Human-flesh search engines have been called upon to find corrupt government officials, unpatriotic citizens, journalists who advocate a moderate position on Tibet, a woman who stomped a kitten to death in stilettos, and cheating spouses.

Disinhibition online can also move to disinhibition offline. A survey of more than a thousand U.S. kids ten to fifteen years old found that those who said they were the targets of some form of Internet harassment were eight times more likely than their peers to admit to taking a weapon to school in the prior month.[11] There are numerous examples of people committing suicide after online interactions or exposure.

Even if the crowd is virtual, the pain of being hounded or excluded by it is real. Even online, minor social exclusion can lead to a dampened mood. Online experiments that manipulated ostracism in a virtual tossing game found that participants who were ostracized reported that they felt bad and had lost their sense of belonging. Ostracized participants were also more likely to conform afterward on another task.[12] Like bullying in real life, Internet bullying may peak in middle school. In a 2005 telephone survey of fifteen hundred U.S. Internet users aged ten to seventeen, 9 percent had been harassed online in the previous year, and 57 percent of those cases were by people they had met online and did not know in person.[13]

Shame in Machines

Digital technologies have transformed the form and scope of shaming, but a subsequent question that programmers, philosophers, cognitive scientists, and others have been asking since the 1980s is how to use logic to formally characterize emotions in those same technologies. Emotions like shame are calibrated to norms, so the question arises: which programmers get to define the norms in computers? A 2001 paper in the *Journal of Artificial Societies and Social Simulation* claims to be "a first step towards including emotions in the computational study of social norms."[14] More than a decade later, there are further advances, with computer scientists giving talks asking, "What values should we program into robots?" However, there has been a lack of success thus far in operationalizing shame in machines.

Displays of emotion might make us more sympathetic toward machines, as they do toward humans, which is presumably why Firefox uses the "Well, this is embarrassing" message when it can't open a webpage. Recall that shaming is often a cheaper form of punishment that prevents harsher forms of punishment. We don't use physical punishment against machines very often. The Luddite destruction of machines related to textile manufacturing is the famous example. A 2004 study found that about 1 percent of slot machine gamblers wind up getting physically aggressive toward the machines.[15] Most of us have personal anecdotes about mild forms of aggression toward machines, like a friend whose sister scarred their

Nintendo controllers with teeth marks. So the question remains whether showing shame would be useful to computer existence, as it was and still is to humans who attempt to avoid worse punishments.

For now, shame remains an emotion that resides in flesh and blood. Digital technologies, like the communication technologies that preceded them, have enhanced shaming's capacity by expanding the audience and increasing the availability of information. Digital technologies have made shaming more accessible to use by the crowd and complicated the question of how not to employ these technologies in error against ourselves.

8

Shaming in the Attention Economy

> The satisfaction is in giving, not receiving,
> information.
>
> —MICHAEL GOLDHABER, "Attention and Software,"
> *Release 1.0: Esther Dyson's Monthly Report*
> (March 1992)

In December 2004, BBC television invited commentary from Dow Chemical spokesman Jude Finisterra on the anniversary of the Bhopal disaster. Dow had acquired from Union Carbide a pesticide plant in India that two decades earlier had leaked gases that killed thousands of people and likely injured hundreds of thousands more. Finisterra's comments seemed genuine: "Today is a great day for all of us at Dow, and I think for millions of people around the world as well. It's been twenty years since the disaster, and today I'm very, very happy to announce that for the first time, Dow is accepting full responsibility for the Bhopal catastrophe. We have a twelve-billion-dollar plan to finally, at long last, fully compensate the victims—including the one hundred and twenty thousand who may need medical care for their entire lives—and to fully and swiftly remediate the Bhopal plant site."

What the BBC did not know was that Finisterra was a fictitious spokesman—a character brought to life by Jacques Servin of the activist group the Yes Men—"impersonating big-time criminals in order to publicly humiliate them." The BBC had found Finisterra through a fake website the Yes Men had set up about the business ethics of Dow Chemical, where they posted a press release announcing that Dow was formally apologizing and compensating the Bhopal victims. This was how the BBC came to invite Yes Man and faux Dow representative Jude Finisterra on for an interview, thinking he was the real thing. Because it took two hours before the real Dow found out about the hoax, the full interview ran twice before it was retracted. The most cunning part of Finisterra's performance was that it then drew attention to Dow's retraction, which stated, for the record, that Dow was absolutely not going to do the right thing. This stunt got a lot of attention and still does, thanks to the Internet.

Attention is a zero-sum game, and shaming requires attention. Michael Goldhaber, a technology writer, was the first to identify the attention economy, in the 1980s. In a 1997 article in *Wired,* Goldhaber further explained, "By definition, economics is the study of how a society uses its scarce resources. And information is not scarce—especially on the Net, where it is not only abundant, but overflowing. We are drowning in information . . . So a key question arises: Is there something else that flows through cyberspace, something that is scarce and desirable? There is. No one would put anything on the Internet without the hope of obtaining some. It's called attention. And the economy of attention—not information—is the natural economy of cyberspace." Goldhaber also presciently identified three problems with the attention economy: "1) The danger of huge inequality between stars and fans; 2) the possibility that increasing demand for our limited attention will keep us from reflecting, or thinking deeply (let alone enjoying leisure); 3) the possibility that we will be so engrossed by efforts to capture our attention that we will shortchange those around us, especially children."

Although Goldhaber focused on cyberspace, many academics place the origins of the attention economy much earlier. Richard Lanham, UCLA English professor emeritus, believes some of the first people to understand the attention economy were visual artists like Marcel Duchamp, who fabricated "a series of attention games with the art-loving public."[1] Shaming's attempts to get noticed are even more complicated today, in an environment heavy on information that is more accessible due to digital technologies. The audience is more distracted than ever. The examples compiled in this chapter are special because they represent the types of real-world shaming that have successfully competed for our attention.

Attention Games

In the attention economy, where tactics can quickly become stale, shaming also has to play games. There are direct forms of shaming, like the Dirty Dozen list of the U.S. politicians with the worst environmental records, and Sam LaBudde's footage of dolphins, and the usual breaking headlines about the bad deeds of banks. Then there are the more theatrical or raucous forms of shaming, like

the Occupy Wall Street movement. It can be hard to mus-
ter attention for the latest boycotts, petitions, or "sucks"
sites (a form of shaming that already seems to be tired).
We want something newer, more fun, more attention-
grabbing. Humor is probably the main reason for the success of ironic shaming, like *The Daily Show* and *The Onion,*
and there is the even greater delight of shaming with the
concealed irony found in the "subvertisements" in the
ad-free magazine *Adbusters* (cigarette advertisements fea-
turing Joe Chemo instead of Joe Camel) or in Jude Finisterra's performance as a Dow representative.

One week after the September 11 attacks, Graydon
Carter, editor of *Vanity Fair* and cofounder of *Spy* magazine, wrote, "I think it's the end of the age of irony." Carter
was, of course, wrong: irony wasn't dead. If anything, irony's cachet is only increasing in the attention economy.
European studies professor Alex Callinicos described
postmodern irony as "the knowing and detached appro-
priation of experiences by an elite that regards itself as
too sophisticated for simple pleasures and unqualified
commitments."[2] Elite and sophisticated, perhaps, but also
operating under a collective attention deficit disorder.
Postmodern irony might just be useful for an audience
that has many different options for where to spend its
time. We continue to want the court jesters to mock the
royalty, but today the powerful can be more difficult to
identify, and the audience is far more distracted.

Concealed irony is a strong form of attention getting,
and various movements have used concealed irony for
shaming, such as détournement and culture jamming.
The Yes Men are a textbook example of using concealed
irony for shaming, and continue to successfully imper-
sonate powerful organizations, including, in 2009, pos-

ing as a U.S. Chamber of Commerce spokesperson and announcing that Chamber of Commerce was reversing its position on climate change policy and going to stop lobbying against efforts to reduce carbon emissions (the U.S. Chamber of Commerce attempted to sue the Yes Men but dropped the case in 2013). On April Fools' Day 2014, National Public Radio (NPR) published a story with the headline WHY DOESN'T AMERICA READ ANYMORE? and linked to it on Facebook. The text of the story was nothing but a request for "genuine readers" not to comment on the piece, because they "get the sense that some people are commenting on NPR stories that they haven't actually read." Commenters then lined up and rebutted the nonexistent story with statements like "it's disrespectful to intelligent Americans to state as fact that America no longer reads."

I expect to see more forms of concealed irony emerge in the twenty-first century, because it's so playful and because it asks more of the crowd, since, to fully appreciate the act, you have to understand the backstory. The audience must understand that *The Colbert Report* is a parody of right-wing pundit shows like *The O'Reilly Factor* to fully admire the performance. The question is whether the human attention span could handle another layer beyond concealed irony. Could someone play Stephen Colbert playing Bill O'Reilly? Nicholas Lemann wondered something along these lines when he wrote his 2006 profile of O'Reilly in *The New Yorker*: "O'Reilly has been playing O'Reilly so successfully for so long, and has developed such a substantial library of hooks, tics, and subplots, that he sometimes seems to be parodying himself, or parodying Colbert's parody of him."

Another artist to play attention-loving games with a tactical twist of shame is the British artist Banksy. The zenith of Banksy's use of concealed irony is his 2010 "doc-umentary" *Exit through the Gift Shop*—if you believe that the main character of the film, Mr. Brainwash, is a Banksy creation and not, as the film would lead you to believe, a full-fledged artist. Roger Ebert began his review of the film on his blog with: "The widespread speculation that [the film] is a hoax only adds to its fascination." If you are in on the ruse (and you believe it is in fact a ruse) that Mr. Brainwash is actually a Banksy creation, *Exit through the Gift Shop* becomes a meta-performance that shames the commercial art world.

This is not to say that concealed irony killed the other direct or literal forms of shaming. There continue to be direct action and real-life protests, like the climate change vigils and the fifteen-month-long protest outside a fur store in Portland, Oregon (in 2005 the owner was fined $40,000 for selling coats made from jaguars, leopards, and other endangered species), and the "Cages of Shame" campaign, in China, to stop the practice of draining the gall bladders of moon bears for traditional medicines. British comedian Russell Brand has created an online show, *The Trews* (true news), that often uses shaming in a calculated although straightforward way (e.g., to expose the brash interview techniques at Fox News). Today there is a lot of information asking for our attention, and a large platform for communicating and storing that information. In this context, shaming can more easily cut through the noise if its signal includes a game.

The Sokal Affair

Another person who understood the rules of the attention economy was NYU physics professor Alan Sokal. In

1996, the academic journal *Social Text* published Sokal's article praising the work of certain postmodernists, titled, "Transgressing the Boundaries: Towards a Transformative Hermeneutics of Quantum Gravity," in which Sokal argued that scientific knowledge was a social and linguistic construct.

Sokal's motivation for writing the article was his growing frustration with an increasing antiscience sentiment from certain subgroups of academics. He had read *Higher Superstition: The Academic Left and Its Quarrels with Science* (1994), by Paul Gross and Norman Levitt, and thought about adding to their criticism of this group of postmodernists, but he imagined that anything directly critical would disappear into "a black hole." So Sokal decided to play a game—something "self-inflicted" that "couldn't be shrugged off"—and instead wrote an article praising "the text of the named authors." He "could be totally playful. The more absurd, the better."

Sokal wrote, "It has thus become increasingly apparent that physical 'reality,' no less than social 'reality,' is at the bottom a social and linguistic construct; that scientific 'knowledge,' far from being objective, reflects and encodes the dominant ideologies and power relations of the culture that produced it; that the truth claims of science are inherently theory-laden and self-referential and consequently, that the discourse of the scientific community, for all its undeniable value, cannot assert a privileged epistemological status with respect to counterhegemonic narratives emanating from dissident or marginalized communities." Phew.

Sokal submitted his thirty-six-page article to *Social Text,* but he is quick to state that the journal was not his main target (though several of Sokal's "named authors" happened to also be editors of *Social Text*). "They were the unlucky ones because they seemed to fit the bill as a kind of trendy journal that just might publish the paper." And publish it is what *Social Text* did, in the spring of 1996. (When I said the idea was ahead of its time, Sokal responded, "Now it's behind the time.")

Sokal then wrote an article that appeared three weeks later in an American magazine focused on academia, *Lingua Franca*, explaining the whole thing was a hoax. He did not issue a press release or get media relations involved. Sokal thought that, in the best case, the whole thing "would land on page 10 of the *Chronicle of Higher Education*." But NPR did a story about the magazine article, and the hoax eventually made its way to the front page of the *New York Times*. The news took a while to cross the Atlantic, but given that several of the academic targets were French, the story eventually found its way to the front page of France's *Le Monde*. Sokal later co-authored a book about postmodern philosophers' abuses of math and physics—*Fashionable Nonsense* (1998)—which played a role in causing this certain science-abusing subgroup of academics to become "out of fashion."

The substance of the Sokal affair is interesting, but the style and strategy are even more so. The game Sokal played was a clever use of concealed irony for shaming. But when I told Sokal this, he said he did not think of it as shame and preferred to think of the affair as self-inflicted embarrassment. "When I did it, I didn't think a whole lot about the psychology. I thought about the sociology." Sokal said that "*shame* was not a word that crossed [his] mind." In any case, it's clear Sokal understood how to use concealed irony.

Inflatable rats

As discussed in the last chapter, shaming has gotten less physical over the past couple of centuries, in part because our discomfort with these forms of shaming has increased: no more public hangings, hot-iron brandings, or prison stripes. It can also be difficult to draw a crowd for protests. In this context, we can expect shaming to continue to become more abstract, because abstract forms of shaming are more acceptable, as well as less expensive. Greenpeace need not gather hundreds of volunteers outside Costco to protest their seafood policies—they can simply fly a blimp with their message above the Costco headquarters and then post the photos on the Internet. But they still need that initial physical act.

Union disputes are another area in which the shaming is less physical now than it was historically. Union confrontations with employers or strikebreakers used to always be face-to-face, sometimes intensely so, and often included picketing and lines to dissuade other workers from entering the site. Pickets require the fortitude of dozens or hundreds of union workers, and the question became whether there was a cheaper way to get attention.

Enter the giant inflatable rat. On any given day, about a dozen are puffed up and standing guard outside various buildings around New York City, as a symbol of a rift between an employer and a union. (The rat was chosen because unions call an employer who does not use union labor a rat.) Additional characters of malfeasance have been added to the stage, including cockroaches, fat pigs, and bedbugs (especially useful for hotels). These union balloons have become an easy-to-spot signal that a dispute has come to a head and an easy way to get attention.

The origins of *Rattus inflatablus* can be traced to the early 1990s, used first by bricklayers in Chicago, which is also, by no coincidence, home to the company that manufactures the union balloons. The balloon company's most popular model is the twelve-foot rat, priced at $4,350, which migrated to New York City in the mid-1990s. The building trade unions in New York were, according to CUNY professor of labor studies and former union organizer and political director Ed Ott, under a lot of pressure at that time, with an influx of cheap non-union labor. And, he says, "employers were undermining standards that had taken a century and a half to establish."

Over time, the rats have gained fame. Locations of New York City rats are reported each day on radio station Q104.3. They have been stabbed, deflated, and taken into police custody. An inflatable rat was even featured on an episode of the television series *The Sopranos*.

Ott explained that unions are "not going to use it for minor infractions—they definitely have a strong tactical sense about the rat." Jack Kittle, political director for District Council 9 painters' union, said, "It's a last resort after we've tried everything we can."

Chicago and New York remain the biggest markets for inflatable rats, but the balloon company said it sells them all over the United States and recently sold one in Florida. Legally, the union balloons are also protected. In 2011, the National Labor Relations Board ruled that the large inflatable rat balloon is not coercive and so does not violate U.S. labor laws. The balloons constitute "symbolic speech." So expect to see more of the rat. "It's a great nonviolent tool that leaves people options," Ott said. "It's good street theater," said Kittle.

The rats also had the unexpected effect of strength-

ening the community itself. "I'm not really expecting sympathy from someone leaving a four-thousand-dollar-a-month apartment on the Upper East Side," said Kittle. "But both residents and workers come out and want to know what they can do to help. New York is really a union-friendly town compared to the rest of the country." Teachers, nurses, and food workers also borrow the painters' union rat from time to time. (For free? "Yes, for free," said Kittle. "We sisters and brothers stick together.")

Neither Ott nor Kittle could explain why other social movements have not caught on to antics like the union balloons, although Kittle said he had heard of an instance in which an employer put up an inflatable cat next to an inflatable union rat. But we can expect the inflatable rats to be used in different ways in the future. "Some people have gotten used to the rat," said Mike Hellstrom, of New York's Laborers Local Union 108, "and part of our strategy is figuring out how to resharpen it to keep it working for us."

Getting a Big Bank's Attention

Just as a bad apple can ruin cooperation in groups, social scientists argue that one or two bad-apple houses can negatively affect a neighborhood, in what is referred to as the broken windows theory. That's why Robert Roberts, a newly retired firefighter who often adds the phrase "boom-boom-boom" to sentences, was upset when the house two doors down in his South Buffalo, New York, neighborhood fell into disrepair and the bank that owned it didn't seem to mind.

"It's really a very sad story," Roberts told me. "The husband had a stroke and lost his job and they had to foreclose on the house." Bank of America then took over as owners, and the upkeep on the house stopped. "I was cutting my grass—I'm not a big dandelion killer or anything, but I want the yard to look nice—and looked at the house, and the grass is two and a half feet high, and thought: That's shitty," said Roberts. Two weeks earlier, his wife had called Bank of America's complaint hotline and waited forty-five minutes before she finally spoke to someone in Texas, but nothing was resolved.

Roberts thought a face-to-face visit would fix things. "I told the guy at the bank that the neighbors wanted to picket the bank, but I told them to wait until the kids were out of school so that they could join," Roberts said. "He told me some stuff, basically passing the buck, boom-boom-boom. He said they'd have the grass cut by Memorial Day." Eventually someone did cut the grass, but Roberts came to find out that the city's guy had cut the grass, not the bank's guy. As a fireman who had recently retired and "watched [his] pension take a dive as we bailed out the banks," Roberts had had enough of banks.

In 2011, after about a year of people trying unsuccessfully with phone calls and face-to-face confrontation, Robert Roberts posed for cameras with a sign that read, BANK OWNED BLIGHT. BROUGHT TO YOU BY BANK OF AMERICA. The photo was on the front page of a Monday edition of the *Buffalo News* and made it into national news outlets, too. The attack on its reputation meant more than the face-to-face visit and finally got the bank's attention. The bank quickly apologized and sent in contractors to cut the grass, trim bushes, repair the gutters, and replace broken doors and windows, to the satisfaction of Roberts and his neighbors.

This case of shaming also shows the difference between

engaging with an individual and with a corporation. I asked Roberts whether he would have put up a similar sign if his neighbor (who had had a stroke) and not the bank had been the one to allow his house to become dilapidated. (I felt guilty even asking the question.) Roberts was indignant. He would have *never* put the sign in the yard of his neighbor. "I'll tell you what would happen," Roberts said. "If the person was living there, his family or his neighbors would have helped him out." We might have both stricter moral standards for individuals as well as stricter rules about how we use shaming against them.

The photo was of Roberts, but he insisted that the tactic should be shared with the whole neighborhood and his local city council member. "I think everyone in the neighborhood was concerned," said Roberts, "but my wife and I are both retired, so we have time to do things like walk over to the bank." Retirees like Roberts are a largely untapped source of shaming.

Again, part of Roberts's success could be attributed to his photo having been posted online (where I—thousands of miles from Buffalo at the time—saw it). Like newspapers and the telephone, the Internet has lowered the cost of sharing information and has further expanded the audience for shaming's maneuvers. But it is important to note that Roberts still chose to engage in a real, physical way. (He did not create a website or a petition.) The Internet is not going to replace real-life action, but it will help spread the news, in this case to Bank of America's decision makers. In one neighborhood in Buffalo, New York, negative publicity worked to change a big bank's behavior, but shaming has many other possible outcomes.

9

Reactions to Shaming

Quand les hommes ne peuvent changer les choses,
ils changent les mots.

(When men cannot change things, they change
words.)

—JEAN JAURÈS, speech at the International Socialist
Congress, Paris (1900)

After the RMS *Titanic* hit an iceberg in the North Atlantic near midnight, there were countless acts of bravery. The wealthiest passenger on board, John Jacob Astor IV, helped his wife and her maid into a lifeboat and then smoked a cigarette as he and the ocean liner sank. The eight-member string band famously played on. Captain Smith, as captains do, went down with the ship.

Then there were the acts of shamelessness. In Lifeboat No. 1, Lady Duff-Gordon was reported to have remarked to her maid how it was a pity that her new nightdress would be lost, and Lord Duff-Gordon, who had married his match, offered the crew members £5 each to refrain from rowing their under-capacity lifeboat back toward the sinking ship (and they accepted).

But perhaps the greatest display of shamelessness was claimed by J. Bruce Ismay, the *Titanic*'s owner, who was on board that fateful night in 1912. First of all, there were too few lifeboats—they could handle but half of the 2,208 passengers and crew on board—and the shortage was a result of a decision Ismay himself had made before the *Titanic*'s maiden voyage. In the end, only 705 people were saved (nearly half of them men), one of whom was J. Bruce Ismay (in collapsible Lifeboat C). Then there were the rumors that Ismay had been the one to tell Captain Smith to speed up the ship. (The ship's speed is assumed to be part of why the iceberg did so much damage upon collision.) The day after the *Titanic* went down, the *New York Times* headline included the words "Probably 125 perish; Ismay safe." Only the second bit was true—more than fifteen hundred passengers had died.

On landing in New York, Ismay was subjected to official inquiry and testified that he felt he had done nothing wrong. But his behavior post-rescue suggested otherwise. Safely aboard the RMS *Carpathia*, which picked up the surviving passengers (all but one of the survivors were in the *Titanic*'s lifeboats), Ismay immediately locked himself away in the cabin that had been occupied by the ship's doctor, where he stayed until the ship docked three days later at Manhattan's Pier 34. In all the cables he sent from the *Carpathia*, he reversed the spelling of his last name both when signing them and talking about himself. These cables were read at the New York court investigation, and Joseph Conrad, who attended some of the hearings, began referring to Ismay as "the luckless 'Yamsi.'"

When Ismay returned to England, he was unwelcome

in British society. Children chanted "Coward, coward, coward" as they passed his gate. He eventually escaped to a remote part of Ireland, where, as one Oklahoma newspaper described it, he hid "in misery and shame." Ismay was never found guilty of breaking the law, but he had certainly defied a code. "To live in isolation is an appropriate penance for someone who has committed a breach of faith with the community of mankind," wrote Frances Wilson in *How to Survive the Titanic, or The Sinking of J. Bruce Ismay* (2011).[1] Ismay would return to London and Liverpool, but he remained withdrawn from their societies—only hosting dinners at home and never eating in a party larger than eight. He didn't report feeling guilt or shame, but Ismay was nevertheless shamed.

Ismay, like many *Titanic* survivors, paid a price for surviving. His particular debt was meted out in public disapproval, which Ismay never confronted. He hid from the other survivors on the *Carpathia* and later from the general public. If shaming causes no discomfort, there is no reason to change one's behavior. But another concern is that shaming is so serious and causes such pain that the transgressor would rather live as an outsider or would rather not live at all. Shame is a painful emotion, and the outcomes of shaming are sometimes very uncertain. While, in the ideal situation, shaming causes its target to conform to the group's idea of appropriate behavior, or makes others conform for fear of shaming, there are other possible reactions, many of them less than ideal, to the threat or actuality of shame.

Stop Shaming Before It Starts

Sometimes shaming is so acute, so powerful, and so effective that the best way to avoid it is to stop it from happening. This is, of course, easier to accomplish from a position of influence. For centuries, stopping shaming has been the prerogative of the powerful—from monarchs to religions to governments to politicians to corporations. For months, the American Press censored the story on the My Lai massacre, in which U.S. troops killed unarmed Vietnamese women, old men, and children. In 2012, WikiLeaks released evidence that in 2009, Dow Chemical had hired a private intelligence firm to spy on Yes Man Jacques Servin, a.k.a. Jude Finisterra. It was the twenty-fifth anniversary of the Bhopal disaster—five years after the 2004 event at which Servin impersonated a Dow representative and "officially" apologized on behalf of the chemical giant for the Bhopal disaster—and Dow did not want to deal with the fallout of another shaming episode. The company was spying on Servin with the aim of stopping any possible shaming before it started.

A more recent example of trying to prevent shaming altogether is the 2006 Animal Enterprise Terrorism Act, supported by Big Agriculture and signed into U.S. law by President George W. Bush. This law extended the scope of terrorism to include activity "for the purpose of damaging or interfering with the operations of an animal enterprise" and was a federal attempt to outlaw whistleblowing on farms and thereby prevent negative exposure. Current laws protecting animals in agriculture are weak, even though the majority of Americans support the humane treatment of all animals, including the ones we eat, which makes undercover footage of farm-animal abuse powerful. In the past few years, leaked videos of workers beating horses, punching pigs, and abusing birds have all led to prosecution. The agriculture industry has a strong interest in avoiding being shamed (or punished in other ways), and several states have also proposed or passed bills, underwritten by the agriculture industry, that prohibit making covert recordings at farms or applying for a job without disclosing ties to animal rights groups, or require that any discovery of wrongdoing be reported within twelve hours. In 2012, Utah passed one of the first laws making it illegal to record video or audio at a farm facility even without trespassing (such as when a woman filmed an injured cow from the side of the road).

Diluting Shame

Another way to avoid the threat or experience of shame is by keeping a low profile, which keeps an individual or a company under shame's radar. In her 2010 *New Yorker* article "Covert Operations," Jane Mayer wrote about the billionaire Koch brothers (owners of Koch Industries): "The Kochs have long depended on the public's not knowing all the details about them. They have been content to operate what David Koch has called 'the largest company that you've never heard of.'" Anonymity is an obvious way to escape shame, because shaming needs a reputation to expose.

In small groups, an individual can build a reputation more quickly. But as the group gets bigger, one act of defection is more diffused and weaker. In bigger groups, defectors are also more likely to find fellow defectors, which helps to normalize defection. (Perhaps this is why BP tried to bring Halliburton in on the Deepwater Horizon blame.) For the reason of increased anonymity in larger groups, shaming can also be weaker in big groups.

Experiments that threatened social exclusion of one person from a four-person group led to increased cooperation, but when the rules allowed for excluding two people from an eight-person group, cooperation did not increase. (Who left the group was decided by a group vote.)[2] This means that part of shaming's success in the public goods experiment my colleagues and I ran could be related to our choice of small, six-player groups.

Today's world is full of one-off interactions and amorphous identities that also make escaping shame easier. When you know you are unlikely to run into the same individuals again, there is less incentive to change your behavior (which is why we did our experiments with students who all came from the same class, and early in the semester, so they were guaranteed to see one another in class again).

There are also so many companies, people, products, and labels that it's impossible for us to keep perfect track of what's going on. Enron, which in 2001 filed one of the largest bankruptcies in U.S. history, hid billions of dollars in debt in hundreds of shell firms, which bought poorly performing Enron stocks so that the company could create a fraudulent company profile and mislead its auditors. Lehman Brothers, in the years before its 2008 collapse, used a smaller firm called Hudson Castle (of which it owned 25 percent) to shift risky investments off its books so that Hudson Castle, not Lehman Brothers, could absorb the "headline risk." The Marine Stewardship Council, the eco-label for fish, uses third-party certifiers that nobody has ever heard of to do the actual certification, which means the MSC can divert the flak of bad decisions. Even high school students are now creating multiple social media accounts—ones where they can be themselves and another one under their real name, where they communicate only their best behavior for college admissions officers.

Another way to dilute shame is to compensate for bad behavior in one domain by building up your reputation in another, the way that Amazon, after a lot of negative publicity for bad behavior such as not collecting sales tax and using international tax havens, launched a charity site called AmazonSmile. The reverse is also true—sometimes reputation can be built up in one domain so that you can afford to behave badly in another. (Think Woody Allen.) Bluestreak cleaner wrasses also know how to do this. These wrasses eat parasites, along with dead or infected tissue, off reef fishes in more than two thousand interactions each day. They are tempted to eat more than just the parasites, but if the reef fish loses too much flesh in the deal, it will refuse to be serviced by the wrasse. When biologist Redouan Bshary watched cleaner fish in the Red Sea in 1999, he noticed that other reef fish watch cleaner wrasses to see whether they cooperate with current clients and avoid the wrasses they see biting off more than they should chew. Some cleaner wrasses are sneaky, though—they know when they are being watched and will build their reputation by politely cleaning small reef fish, allowing the big ones to observe them on their best behavior with the small fry. When the big reef fish comes in for cleaning, these wrasses will then cheat and eat some of the big reef fish's flesh along with its parasites, fattening themselves on their defection.[3]

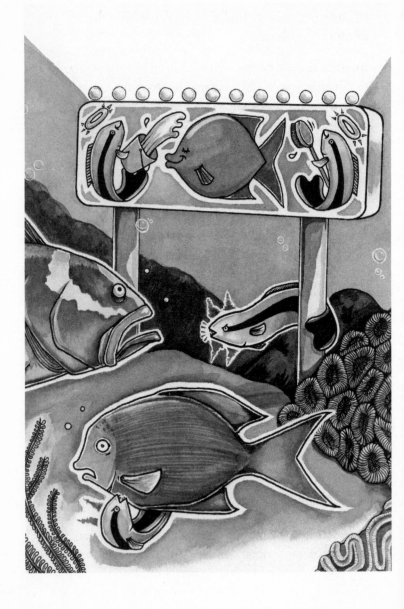

Disabling Shame Through Different Mental States or Moral Categories

Another way to escape shame is to have no aversion to the prospect of social disapproval from the start. This strategy seems increasingly prevalent in cultures that have become more individualistic. Closeness to others is an important antecedent of shame, so distance from others—physical as well as mental—means that shaming is less likely to have an effect.

We appear to have stricter moral standards for individuals than for groups, which might serve to insulate some groups from shaming. But not all groups are created equal. People see McDonald's, the U.S. Marine Corps, and Hare Krishnas as having high "group cohesion" and "group mind"—meaning that the group has intentions and makes plans—but people do not perceive the same cohesion in groups of "blondes" or "golfers." They also perceive certain groups (ones with high "group mind") as being more responsible for their collective actions: Citibank and the U.S. Navy are seen as more responsible for what they do than are car owners and tennis players.[4]

At the same time, we allow certain groups of people to behave in ways that we would never allow an individual person to behave. Corporations, for example, are allowed to operate for the overriding mission of profit and are perhaps held to a different standard of behavior. Milton Friedman explained that a corporate executive may have many other responsibilities, such as "to his family, to his conscience, his feelings of charity," but that "there is one and only one social responsibility of business—to use its resources and engage in activities designed to increase its profits so long as it stays within the rules of the game, which is to say, engages in open and free competition without deception or fraud."[5] According to Friedman, it can be immoral for a corporation to be moral. These views are part of the reason Friedman took a cream pie to the face during a 1998 conference on the privatization of public education.

Corporations and other groups are not the only ones evading shame (again, perhaps for adhering to a different set of norms). We also seem to have looser moral standards for machines. The ultimatum game is an experimental tool used to study fairness. Money is given to one player who can offer his partner all, some, or none of it. The partner can then choose to accept the offer, in which case both players walk away with the proposed split, or the partner can reject the offer, and neither of the pair receives any money. The most common offer is an even split, and people around the world routinely reject offers below 20 percent, which means that neither person receives any money. But when humans thought they were getting an offer generated by a computer, their decisions changed. Participants rejected unfair splits of 80/20 or 90/10 from a supposedly human partner (shown in a photograph) but accepted those same unfair offers when they thought they were being made by a computer. Brain areas that showed more activation during unfair offers from humans showed *less* activation when computers made the exact same unfair offers.[6] A similar experiment that tested skin conductance—a proxy for psychological arousal, based on the connection between sweat glands and the sympathetic nervous system—found higher conductance during unfair offers, but only when those offers were made by humans, not computers.[7] This means that our morality is context- or agent-dependent, which is worrisome in an era of drones and high-frequency trading (automated stock trading now accounts for the majority of market activity). An additional concern is that if groups and machines lack the emotions that would fear shame or lack the apparatuses for morality, they might be functionally less susceptible to shaming anyway.

Escaping Shame After It Starts

Trying to stop the shaming after it has started is trickier than stopping it before it begins. Lawsuits tend to only call more attention to bad publicity. One way to escape shaming is to simply wait it out, and this is standard corporate protocol for bad publicity. Time heals all wounds.

Another tactic to stop the shaming is to undermine the credibility of its source with a counterattack. After the backlash against the Komen foundation's decision to stop funding Planned Parenthood (which they then reversed), former Komen executive Karen Handel (in charge at the time of the kerfuffle) responded to the negative publicity with the book *Planned Bullyhood* (2012). In October 2010, WikiLeaks founder Julian Assange walked out of a CNN interview because the interviewer brought up the rape charges against him. In a subsequent interview with CNN's Larry King, Assange argued that the charges against him were disproportionate to the deaths of 109,000 people that had been revealed by government documents, and tried to turn the tables, adding, "CNN should be ashamed."

Sometimes the best way to escape shame is to literally escape and hide from the crowd, as Bruce Ismay did. We already know that humans physically shrink when confronted with a shameful experience. In Mandarin, one word for shame, *diu lian,* translates as "loss of face"—an unsurprising turn of phrase, given that disfigurements were one of the earliest causes of shame.[8] The concept of "face" has its origins in Chinese culture and behavior, although "saving face" is a common expression across many cultures, with both literal and metaphorical reputational meanings. When high-ranking members of the native Pacific Northwest Tlingit society were injured, particularly on their face, they would stay inside until healed.[9] (Today the same can be said for post-op plastic surgery patients.) Online, when people find themselves the subject of shame due to a tweet, they will delete their Twitter accounts or make them private—a form of digital hiding.

But it's not just the primary victims of shaming who might choose to hide. Shame can also be contagious, extending to friends, family, and other members of the group and likewise besmirching their reputations. Children of hoarders talk about "doorbell dread"—the fear that someone will discover the collection their parents have amassed. After a twenty-four-year-old gunman killed moviegoers in Colorado, his parents went into hiding. (They emerged for the arraignment.) In the past in Japan, where traditionally so much of shame was a family matter, ritual suicides would allow a high-ranking family to be freed from shame.

It can be difficult for individuals shamed by a family member or religious leader to cut ties to escape shaming, but groups, especially corporations, do it all the time. BP announced it would fire CEO Tony Hayward three months after the oil spill in the Gulf of Mexico began, and the Komen foundation installed a new CEO after negative publicity for its Planned Parenthood decision. For celebrities whose shame spills over to their corporate sponsors, escaping shame means ending contracts. Accenture, AT&T, Gatorade, and General Motors dropped Tiger Woods after his extramarital affairs became public. H&M, Chanel, and Burberry ended Kate Moss's modeling contracts after the press published photographs of her snorting white powder. (Her work has since picked up again, giving a nod to the wait-it-out strategy.) The Food Network, Walmart, and Random House dropped celebrity cook Paula Deen after her racist remarks. Fashion brand Christian Dior fired designer John Galliano following the posting of a video of his anti-Semitic rant.

Shaming can also be escaped by cutting ties to a place and moving to a new locale where either the cultural norms are different or you have yet to be detected as a miscreant. The hog-farming industry in North Carolina got a heap of negative publicity for documented cases of animal cruelty and for the environmental impacts of manure spills resulting from Hurricane Floyd, in 1999. To escape the shaming, the industry moved many of the hog farms to Iowa, where operations could resume as before without the same levels of scrutiny.

Those attempting to escape shame can involve hiding not oneself, but the evidence. Owners of private planes who have gotten bad press for flying them, like executives at General Motors, can, under a new exemption, block the public's ability to track the jets (the independent investigative journalist group ProPublica has done a great job tracking this story).[10]

Cutting ties with a name is another option to escape from shaming, as the luckless Yamsi did. After negative exposure, U.S. federal prisons now refer to solitary confinement as the Special Housing Unit, and the Department of Defense's School of the Americas is today renamed the

Western Hemisphere Institute for Security Cooperation. The Obama administration shunned the Bush administration's "global war on terror" in favor of the vague "overseas contingency operation." After negative publicity for its performance in Iraq, Blackwater changed its name to Xe in 2009 and, in 2011, to the more pedantic Academi (although the online retail arm of Blackwater, which includes embroidered "cotton tactical shirts, performance polos, response jackets, and integrity hats," is still, as of this writing, operational, suggesting that Blackwater's reputation was mixed). BP was rumored to have considered changing its name after the Gulf spill but decided it had spent too much building the brand to get rid of it.

Even consumer products adopt new names to avoid shame. Pizza's growing reputation as junk food led Pizza Hut to attempt the new brand "The Hut," or, in some places, "Pasta Hut"—but the response was so sardonic that the brand recommitted to the name Pizza Hut. Since 2009, retailers have sold Aspartame under the name AminoSweet. In September 2010, the Corn Refiners Association petitioned the FDA to permit high-fructose corn syrup to be called "corn sugar." (The FDA denied the request.) Some fish were once given the names slimehead, Patagonian toothfish, and spiny dogfish, before they were common on our dinner plates, but now that so many species are overfished, these fish head to market under the new, A-list names orange roughy, Chilean sea bass, and rock salmon.[11]

Sometimes it's not just names that are changed, but a person's or a company's entire reputation. Journalist Graeme Wood exposed a chicanery-filled tale of reputation management in a 2013 article in *New York* magazine.[12] Wood had heard that a guy he knew in college, P.U.

(abbreviated here to anonymize him), was being charged for federal tax fraud and for helping his mom (!) smuggle hundreds of thousands in cash out of Zurich. His mother was found guilty, but charges against him were eventually dropped. Yet an online search of P.U.'s name showed his online reputation had been pooched. (P.U.'s mom was not just responsible for her son's run-in with the law—she had also given him a unique first name that made him particularly susceptible to the search engine's spotlight.)

Out of curiosity (and perhaps a little schadenfreude), Wood set up an online alert for P.U., and over the months after the charges he received a series of alerts for some strangely positive pieces (thirty-three unique sites in total). Wood eventually sleuthed out that P.U. had hired the reputation-management firm Metal Rabbit Media—"a boutique shop for the online reputations of very wealthy people." (Its services can cost $10,000 a month.) Metal Rabbit did not just create positive media; one of its services included creating a fictitious do-gooder with the same name as its client—"doppelgängers . . . so that one could never be sure whether the snark on the web referred to the fakes or to the misbehaving real McCoys. [Wood] imagined a future in which rich people create dozens of scapegoats for themselves, like Saddam Hussein with his body doubles, and wondered how some data-mining bot might tell the difference."

This is another divide within the digital divide: there will be those who can afford to remove the stains on their digital reputations and those who cannot. And industries will continue to profit off of reputation management and public relations. There are websites that post Florida mug shots (the files are made available on the basis of a request for public information) and other sites, in cahoots with the mug-shot sites, that offer, for a fee, to take them down. Paying to take content off the Web is a growing market.

If this is what individuals are up to in terms of reputation management, we can only imagine what companies are doing to try to scrub any stain from their brand. "Astroturfing" is the term for gossip that appears to originate from a small grassroots organization or an individual (for example, a mommy blog) but is actually funded by big business or political interests. The film *Truthland,* a rebuttal to the *Gasland* documentary, about the negative effects of natural gas extraction through hydraulic fracturing ("fracking"), depicted a Pennsylvania "teacher, dairy farmer, and mom" who planned a road trip to find out the truth about fracking. Actually, the film was conceived and later promoted by the public-relations arm of the oil-and-gas industry. Academic research can also help companies avoid shame. Authors of a 2012 study in the journal *Marketing Science* recommended that companies monitor online gossip, focus on the volume of negative gossip, and "use more off-line television advertising because it increases the volume of online chatter while decreasing negative chatter."[13]

Other Liabilities

One of shaming's biggest liabilities is that it could further isolate, rather than reintegrate, offenders. The scarlet letter "had the effect of a spell, taking [Hester Prynne] out of the ordinary relations with humanity, and enclosing her in a sphere by herself." This isolation could lead a transgressor to find it easier and more appealing to live his or her life as a transgressor than as a member of the group.

In our public goods experiment investigating honor

and shame, everyone played two additional rounds after the most or least generous players were exposed. On average, the two least cooperative and shamed players further decreased their cooperation in rounds 11 and 12, while the most cooperative and honored players continued to cooperate and donate to the public good, suggesting that participants acted in a way that reinforced their reputation.

The overuse or misuse of shame or the apparent prevalence of shameful behavior can even reshape social norms. "Such helpfulness was found in her . . . that many people refused to interpret the scarlet A by its original signification," wrote Hawthorne. "They said that it meant Abel; so strong was Hester Prynne, with a woman's strength." In states that require brightly colored license plates on cars registered to convicted drunk drivers, the plates are seen by some as a source of pride and referred to as "party plates."

Another potential problem is that exposing all the slumlords or polluters could communicate that these behaviors are common and normalize the behavior. This has been called the boomerang effect—the tendency for people who had abstained from an undesirable behavior to begin to partake because they come to incorrectly perceive that behavior as normal. One way to avoid this is to remind the audience of the norm, just as the California tax-delinquent website reminds people that "nearly 90 percent of taxpayers pay the taxes they owe."

Confronting Shame

For some, it is easier to endure shame because the financial benefits of their behavior simply outweigh the cost of being shamed. "Those people who are most likely to defy social norms and risking shaming sanctions, even within close-knit societies, are the very rich and the very poor," wrote Toni Massaro.[14] The rich are "insulated by their wealth," and the poor have "less 'social standing' to lose." In other words, there are the people who do not conform to social norms because they cannot afford to care, and there are the people who do not conform because they can afford not to care. Dostoyevsky's Raskolnikov was convinced he was a "great man" and therefore could commit a crime without receiving the punishment, which it turned out he could not, and he was (spoiler alert) convicted of murder and exiled to Siberia. This links back to not feeling part of the group and instead feeling above or beyond it.

For those looking to actually quell the shaming, one option is to express gratitude or remorse. After AIG's negative publicity for spending some of the bailout money on lavish retreats, the company decided to make some publicity of its own. AIG's Thank You America campaign was an expression of gratitude for the federal bailout money and probably would have at least somewhat ingratiated AIG to the public, improved its reputation, and alleviated some of the shame from the financial crisis. But one week later, AIG considered joining a lawsuit against the government because it had missed out on some profit due to restrictions during the bailout—a bailout that came about after a financial crisis that AIG antics helped precipitate. The headlines were not flattering. The *New York Times* said, RESCUED BY A BAILOUT, AIG MAY SUE ITS SAVIOR. *Gothamist* was more blunt: AIG: THANK YOU AND F*CK YOU, AMERICA. AIG's board decided the next day against joining the litigation, although AIG's former chief executive, with help from others on Wall Street, later filed a lawsuit.

Another way to make amends is by apologizing. Apologies are part of the post-shaming healing process—both for the transgressor and for the crowd. In 2010, James Frey added to his 2003 memoir *A Million Little Pieces* a three-page author's note apologizing for having fabricated large portions of the book—perhaps on account of being shamed by Oprah and the *South Park* spoof "A Million Little Fibers" (in which a towel writes his memoirs and pretends to be a human). But just as it is difficult to know whether people feel shame, it is difficult to know whether an apology is sincere. With some people, it is clear that they knew they were doing something the crowd saw as wrong, and their apology is the result of being caught—as with Clinton's "I don't think there is a fancy way to say that I have sinned" or Tiger Woods's "I knew my actions were wrong, but I convinced myself that normal rules didn't apply."

In the best case, the threat of shame makes people change their behavior to match the group's expectations—although sometimes it's nothing more than a token gesture. In November 2008, after GM, Ford, and Chrysler had lost 25 percent of their market share from the previous decade, the CEOs of the three major U.S. car corporations went to Washington to ask for $25 billion in public funds, and each CEO flew separately in his private jet. During the meetings, members of Congress chided the CEOs for their gall. One said that it was "almost like seeing a guy show up at the soup kitchen in a high hat and tuxedo," and another asked the three to raise their hands if they were willing to sell their jet and fly back commercial (no one did). Almost certainly due to the threat of shame, the three CEOs each drove his company's most fuel efficient car when they returned to the Senate Banking Committee hearings one month later. They all promised to sell their jets (although it's not clear that they did), and eventually Congress agreed to billions in bailout money.

When shame works without destroying anyone's life, when it leads to reform and reintegration rather than fight-or-flight, or, even better, when it acts as a deterrent against bad behavior, shaming is performing optimally. (Whether the norm that the shaming encourages is optimal is, as always, open to debate.) If shame has a goal, it is to deter what the group considers bad behavior.

One could argue that it was shaming that led Microsoft mogul Bill Gates to build the largest charitable foundation in the world and become one of the most respected philanthropists. In the late 1990s, Gates was under a serious spotlight of negative attention after his irritable testimony during the Microsoft antitrust trial and the court's decision that the company was indeed acting as a monopoly. Gates was also hit with several cream pies. In 1999, the William H. Gates Foundation (a name he and his father share) became the Bill & Melinda Gates Foundation, and Gates Jr. doubled its endowment.

Exposure, a key ingredient of shame as well as of honor, was also what Bill Gates himself turned to when he wanted to encourage other billionaires to be more philanthropic. In 2010, Gates and Warren Buffett unveiled the Giving Pledge, a list and website that publicizes the megawealthy who have made a similar (nonbinding) agreement to give the majority of their wealth to charity. Not surprisingly, the media has reported not only who is on the list, but who is conspicuously absent, including Oprah Winfrey and the Walton family (of Wal-Mart Stores). To encourage philanthropy, Gates also appealed to the power of reputation.

Shaming Gates for being a monopolist, shaming fisheries for killing dolphins, and shaming manufacturers for poor working conditions have all led to what the audience deemed better behavior and could therefore be described as effective uses of shaming. Here we find the real power of shame: the fear of it can make individuals or institutions conform to what the group thinks is acceptable behavior. Shame's service is to the group, and when it is used well and at the right time, it can make society better off.

10

The Sweet Spot of Shame

> It's not about what it is. It's about what it can become.
>
> —ONCE-LER, *The Lorax* screenplay (2012)

Before he became mayor of Bogotá, Antanas Mockus was an academic, with master's degrees in mathematics and philosophy, and the president of Colombia's Universidad Nacional. His entrée to government came about after an act of shamelessness. In 1993, needing to get the attention of a rowdy student crowd, Mockus dropped his trousers and mooned the assembly. He got everyone's attention. He also lost his job.

"Shame and shamelessness are recurring themes with Mockus," Felipe Cala told me. Cala was born and raised in Bogotá and wrote a chapter of his dissertation, on culture-based advocacy in Latin America, about Mockus. According to Cala, Mockus "capitalized politically on the shamelessness of his action." Using his notoriety and a total of $8,000—one of the cheapest mayoral campaigns in Colombia's history—Mockus was elected mayor. He took office in January 1995, and his six years in power (1995–1997 and 2001–2003) would prove to be as unconventional as his time on the university stage.

Mockus inherited a tough city filled with political kidnappings, assassinations, and bombings by drug cartels. By the mid-1990s Colombia's homicide rate was five times higher than the Latin American average (which was higher than the rest of the world's). According to a talk Mockus gave at MIT in 2012, Bogotans behaved well in private but believed that "being anonymous was generating bad public behavior." Mockus hoped to make the city safer by encouraging a culture of self-regulation.

One of Mockus's goals was to make Bogotá safer for drivers and pedestrians. He helped distribute more than one million cards that were red on one side and white on the other and encouraged people to use them against drivers to show approval or disapproval (similar to the cards used during soccer games). In July 1995, a survey showed that nearly a quarter of the citizens had one of the cards, and two-thirds had heard about them. Of those who knew about the cards, 70 percent thought they were a good idea in trying to control dangerous driving.

Mockus also hired twenty professional mimes to draw attention to jaywalking, reckless driving, people not wearing seat belts, and excessive honking. The mimes were so popular that the city trained four hundred more, and the mimes are partly credited with the decline in the number of deaths caused by traffic accidents, which fell by 50 percent between 1995 and 2002. According to Cala, the mimes Mockus hired "became one of the trademarks of his time."

In 2001, Mockus sponsored a night for women, on which he requested that the city's men stay home in the evening, either caring for their children or just keeping off the streets. He also introduced a voluntary curfew to cut down on the drinking and violence, and, to seed its success, he canvassed neighborhoods wearing a clock around his neck. When there was a water shortage, Mockus made a public service announcement in hopes of promoting a new norm to conserve water. Standing naked in the shower, he explained how he turned off the water as he soaped, and asked his fellow citizens to do the same; two months later, people were still using less water.

"His M.O. was to get people to reflect," said Cala. Mockus used shaming to expose bad behavior and also humbled himself before the public through his own shamelessness—mooning the audience and being filmed naked in the shower. His most solemn expression of humility was the heart-shaped cutout he put in the bulletproof vest he was often required to wear, which was directly over his own heart—an unpopular move with his security team. Mockus's frequent use of humor and self-deprecation comes up in almost everything written on him. He also knew how to play attention games, and his strategic use of art earned him an invitation to the 2012 Berlin Biennale. Mayor Mockus frequently found the sweet spot of shame.

Finding Shame's Sweet Spot

Some people see shaming as an outmoded tool, useful at some point but no longer needed. Others see shaming as similar to nuclear technology: effective but also potentially dangerous and likely to fall into the wrong hands. For shaming's role in the twenty-first century, we should think less about the traditional forms of shaming—the stocks or the scarlet letter—and more about what shaming might become. Mimes who shamed bad driving in Bogotá led to a decrease in traffic accidents. Shaming tax delinquents led to an increase in tax revenue. Labels that singled out salty foods in Finland significantly decreased salt consumption.

In some cases, like highly salty foods, shaming might be all we need or want as a deterrent. Other times, when our values change more quickly than the institutions designed to enforce them, shaming companies or governments can be the first step toward institutionalizing formal rules and punishment, as it was in the case of banning child labor. Sometimes, shaming is successful in establishing a norm, as it was with decreasing the wearing of fur more than a decade ago, but without formal rules to follow up, the norm can relapse (as happened with fur).

Recall the seven habits of effective shaming: The transgression should (1) concern the audience, (2) deviate widely from desired behavior, and (3) not be expected to be formally punished. The transgressor should (4) be sensitive to the group doing the shaming. And the shaming should (5) come from a respected source, (6) be directed where possible benefits are highest, and (7) be implementated conscientiously.

But the sweet spot of shaming implies more than just effectiveness—it also implies that the shaming is permissible. For the sweet spot of shame, like so many other fickle things, the Goldilocks principle is instructive: shaming should not be too weak or too strong, too brief or too permanent, used too infrequently or too often. To further complicate things, what is considered too extreme is always changing. Shaming must remain relevant to the audience's norms and moral framework. I'd like to speculate about some things that might make shaming more acceptable, at least in a Western or, more specifically, United States context (while acknowledging that this is an area that could use more research).

A reminder, first, that a transgressor is needed. For some transgressions, there is no clear transgressor, such as in the cases of widespread poverty, disease, hunger, or water shortages, which makes shaming difficult to use with any outcome. To get shaming off to an acceptable start, due process is also important. A pure utilitarian might say that shaming is justified if society is better off, even if the shaming was directed at someone who had not transgressed. That pure utilitarian view is likely to be rejected because it violates norms of fairness. Shaming should also work to stigmatize bad practice, rather than focusing overtly on specific people or institutions.

Shaming the bad practices of institutions, companies, or countries is probably not just more effective in terms of bringing about large-scale changes in behavior (consider that, of the one hundred largest economies in the world, half are corporations—Walmart's revenues rival Argentina's GDP), but shaming groups is probably more acceptable. Groups are less prone to emotional suffering in the face of shaming, because groups do not have "human dignity" the way that individuals do. Not all groups are created equal, though, and we might be more comfortable shaming some groups than others.

Acceptable shaming also tends to focus on the powerful over the marginalized. A shaming campaign that further ostracizes the people already most ostracized by society (or that further stigmatizes the most stigmatized) will be less defensible. We typically dislike high-status people picking on low-status people (although we also clearly tolerate it to varying degrees). In the cases of exposing animal cruelty in factory farms, it would be a mistake to direct the shaming at the individual farmworkers.

We have also become more uncomfortable with shaming that directly affects our physical space. Branding, dunce caps, and scarlet letters are no longer acceptable, and it may be said that these forms go beyond shaming and into the realm of humiliation. Judges in certain U.S. states do still require forms of physical shaming, like making people carry signs for a certain amount of time announcing that they shoplifted, but these punishments are controversial. As recently as 2007, a man suspected of selling drugs to children was tarred and feathered in the Northern Ireland city of Belfast—and media coverage suggested widespread disapproval of the act.

Shaming today has greater potential, due to the many possible forms of abstraction that let us avoid physical stigmatization or confrontation, such as the inflatable rats. Perhaps the most important abstract tools shaming must contend with are digital technologies, which have increased the amount of behavior we can keep track of, the scale of the audience, and the indelibility of the information. Online, just as in real life, the form and design of shaming matters, and any policies and platforms that invoke shaming should be designed with prudence.

Although abstract forms of shaming are becoming more popular, physical, face-to-face engagement is not likely to disappear. There is a market for the book *Pie Any Means Necessary: The Biotic Baking Brigade* (2004), about "the fine art of landing a freshly baked delicacy in the face of the reactionary, pompous, and otherwise deserving." We should recognize that the pie of shame is cream-filled—a pecan pie would be potentially harmful. "When you say things with your body, it's stronger than if you say them just by words," Antanas Mockus said during a keynote address. CUNY professor Ed Ott, the former union organizer and inflatable-rat supporter, told me, "People still need to see, feel, and smell each other if they want to know what each person really wants."

Protecting Shaming in Society

Shaming is one of our most successful means of nonviolent resistance and a tool that can be used in the absence of formal punishment, but it also loses its power as it is used more. This is one reason to eliminate, as best we can, the frivolous uses of shame.

We should also ensure effective shaming organizations by recognizing the need for those organizations to be independent from groups that might be their target. Greenpeace maintains that their ability to engage in shaming is rooted in the fact that they do not take corporate or government funding. Revolving doors, board memberships, and charitable donations are all subtle ways of preventing shaming. The documentary titled *Park Avenue: Money, Power, and the American Dream* focused on America's growing economic inequality and the people at the top of it, including billionaire David Koch, of Koch Industries. It aired in 2012 on WNET, a New York public television channel that also, as of 2006, benefited from the board membership of David Koch. Jane Mayer, at *The New Yorker* (chief shamer of all things Koch), discovered that before the film aired, WNET president Neal Shapiro had called Koch as a "courtesy" and allowed the spokesperson for Koch Industries to issue a statement that aired immediately after the film, and called the film "disappointing and divisive."

Polling the Audience

Given that the crowd is asked to be part of the punishment, it makes sense to ask the crowd whether shaming punishments are acceptable. A 2008 article in the *Seattle Times* asked readers whether they agreed with issuing bright yellow license plates to drivers with DUIs, as many other states had done. There were more than five thousand responses, and 45 percent said they agreed with the yellow plates while 52 percent disagreed, and the remainder were undecided.

For twenty-five years, New York City put neon stickers on cars violating street-cleaning rules. Officials insisted that they worked: street cleanliness had improved 30 percent since the stickers were introduced. The stickers also pointed out the collective nature of the dilemma: THIS VEHICLE VIOLATES NYC PARKING REGULATIONS. AS A RESULT, THIS STREET COULD NOT BE PROPERLY CLEANED. A CLEANER NEW YORK IS UP TO YOU. I thought the parking stickers hit the sweet spot of shaming, but in March 2012 the city council unanimously voted to outlaw the stickers, overriding a veto by then Mayor Bloomberg. It would have been interesting to know how other New Yorkers felt.

Three of my colleagues—psychologists Jessica Tracy and David Pizarro and mathematical biologist Christoph Hauert—and I surveyed 111 U.S. residents online about their opinions regarding different shaming scenarios. We wanted to know if it mattered whether the shaming was physical or online. We gave people scenarios like this: "Charles Williams committed income tax fraud and cheated the community out of more than 1 million U.S. dollars. As part of the sentencing, a judge required Williams to put a sign outside his house that described his crime." A later scenario would offer a slight variation, but instead of a sign outside his house, the punishment would be an online posting of his name on a state list of delinquent taxpayers. The only difference between the two texts was whether the delinquent taxpayer's name was on a sign outside his house or on a website.

Overall, across all sixteen scenarios we gave them, subjects found shaming to be fairly acceptable (5.42 on a 1–7 scale where 1 was "very unacceptable" and 7 was "very acceptable"), suggesting that the general public is not strongly opposed to the use of shaming as a punitive measure. Keep in mind that these shaming scenarios were pretty mild, and the crimes involved were pretty serious: stealing $1 million or not paying $1 million in taxes. Respondents also indicated that they found posting someone's name online significantly more acceptable than putting a sign on their lawn, which was interesting, given that our subjects were recruited online and therefore know about the huge online audience. These results suggested that the physical presence of an audience was more problematic than audience size. However, with regard to the Internet, these are still early days. Impressions of what is acceptable online could easily change.

Why Not Transparency?

Given the pain and discomfort shaming can cause and the difficulty in finding shame's sweet spot, why not simply expose everyone's behavior? Transparency is a popular panacea today—promoted as the cure for society's ills. The premise of transparency is that people improve their behavior when everyone can see it, so therefore we should make everyone's behavior visible. Unlike shame or honor policies, which aim to expose only the worst or best slices of a population, transparency policies reveal the behavior of every entity involved, and therefore might at first glance seem more democratic and fair. But contrary to our instincts, a shaming policy can sometimes be both more effective and more acceptable than transparency.

For one thing, transparency might provide too much information. Transparency policies reveal the behavior of every entity involved and often require the audience to determine what is acceptable, which can be difficult. Do we need or want to know how much every company pollutes? Or is a list of the hundred worst corporations sufficient? Do we want to know about every tax delinquent? Might that hurt some people who did not pay their taxes because they were down on their luck and it was a particularly low earning year? In California, delinquents don't make it onto the list for anything less than $250,000. This helps to protect the least significant (in terms of the state's debt) as well as the poorest of the state tax delinquents.

Shaming policies can also threaten exposure and then allow people who change their behavior to avoid it; transparency does not allow for such an option. Like radios, which come with both a seek button, to provide access to all radio frequencies, as well as a scan button, which accesses only the strong frequencies, we need both options. Transparency is often a good tool, but it's not always the best.

Transparency policies can on occasion act almost identically to shaming policies, because people are most interested in the negative information. "Shame Is Not the Solution" was the title of a Bill Gates 2012 *New York Times* op-ed after New York State ruled that all teacher performance assessments could be made public, which is actually a transparency policy, not a shaming one. Gates called the court's decision "a big mistake." He also wrote, "At Microsoft, we created a rigorous personnel system, but we would never have thought about using employee evaluations to embarrass people."

Yet employee evaluations, another form of transparency, have indeed been used at Microsoft and even implicated in the company's failure to innovate over the past ten years. Microsoft used stack ranking, a system in which every few months certain employees are labeled as poor performers and everyone lives in fear of being at the bottom of the stack. If poor performers do not improve, they are asked to leave the company. When Kurt Eichenwald interviewed former Microsoft employees for his 2012 *Vanity Fair* article "Microsoft's Lost Decade," he wrote that every single worker "cited stack ranking as the most destructive process inside of Microsoft, something that drove out untold numbers of employees."

Microsoft inherited stack ranking from the playbook of Jack Welch, CEO of General Electric from 1981 to 2001 and co-author of self-effacing titles like *Jack: Straight from the Gut, Winning: The Ultimate Business How-To Book,* and (the double-coloned) *Winning: The Answers: Confronting 74 of the Toughest Questions in Business Today.* Under Welch's two-decade watch, General Electric grew in market value from $14 billion to $484 billion, and it is because of this that people from Croatia post online photos of themselves holding his book. Welch's stack-ranking system, also referred to as forced ranking, rank-and-yank, or, euphemistically, the "vitality curve," led to competition and caginess at Microsoft and stifled innovation. One can envision social dilemmas in which this harsh form of transparency would be useful, but inventing new software probably isn't one of them. The threat of shame or transparency should be avoided for any creative pursuit (which seem to be better served by reward mechanisms). Negative exposure is a much better tool for promoting cooperation and social cohesion, especially in the absence of other options.

Remember Shame's Social Nature

The more social the transgression, the sweeter the shame, because the audience is inherently interested in the bad behavior. Ultimately, shaming might not be enough to solve the problem, but the greater fallacy, particularly for collective-action problems like climate change and overfishing, is the idea that we can individualize them. In our preoccupation with the spirit of individualism and free-market ideology, we bought in to the notion that the biggest way an individual can make a difference is as a consumer. This has allowed for the ascendancy of guilt, which helps sell products like sustainable seafood, organic foods, and carbon counters. Consumers are swept up in using reusable bags and mugs and turning off the lights. This is like taking vitamin C after fracturing your skull in a car accident: it is not wrong; it is just so far from what is needed to actually fix things. For large-scale cooperative dilemmas, it is not sufficient that a small group of people feel guilty, and it is certainly not enough that this small group engage with that guilt as consumers. We need a tool that can work more quickly and at larger scales.

We live on one planet. As far as we know, we are the

only species that really understands that. And we alone bear the responsibility for balancing human interests and the interests of nonhuman life. Shame is one of many possible tools to help us get along. Who is "us"? Our neighbors? People inhabiting small island states threatened by sea-level rise? Mountain gorillas in Central Africa? Toothfish in the Ross Sea? The Monterey pines of California? Successful species will likely be those that recognize, implicitly or explicitly, life's interdependency.

Given that shaming is inextricably linked to norms, its role in our future is crucially dependent on what norms the future brings. "For one species to mourn the death of another is a new thing under the sun," wrote Aldo Leopold in 1947. "The Cro-Magnon who slew the last mammoth thought only of steaks. The sportsman who spotted the last [passenger] pigeon thought only of his prowess. The sailor who clubbed the last auk thought of nothing at all. But we, who have lost our pigeons, mourn the loss. Had the funeral been ours, the pigeons would hardly have mourned us."[1]

This is our plight in the modern world: we now understand the exact dimensions, mass, and materials of the earth, where it is suspended in the universe, the uniqueness of the life that inhabits it, the problems we have created on it, the solutions we have not implemented, and the solemn reality that this earth is, to date, our only home, and our only chance. The new norms required in this context are big ones, and their formation and reinforcement can be assisted by the wise use of shame.

Appendix

Shame Totem v.2.1

Desiring to create a shaming initiative of my own that reflected some of the seven habits of highly effective shaming, I first called upon the crowd for input. Using an online survey, I asked five hundred anonymous Americans to select the ten firms that had most negatively affected society, from a list of the fifty biggest publicly traded corporations. But I wanted to find a way to expose bad behavior by providing more than just a list of culprits.

For seven years, I lived in British Columbia, Canada, and during a visit to the Haida Heritage Centre, on Haida Gwaii—a set of Canadian islands just south of Alaska—I learned about shame totem poles, which were one of eight different types of poles carved from red or yellow cedar or Sitka spruce trees by almost every single Pacific Northwest tribe from the eighteenth century onward. Because their meaning is difficult to discern without the requisite symbolic points of reference, totem poles have typically been appreciated by settler cultures for their magnificence and beauty. But from the oral histories of native communities, as well as from early ethnographic studies, we know that beauty was not their only point.

Shame totem poles signaled to the community that certain individuals or clans had transgressed. One of the most famous shame totems includes three frogs lined up across a pole, said to represent three women from the Kiks.ádi clan (whose totem is the frog) who were allegedly living with a different clan. When the Kiks.ádi chief was presented with a bill for the keep of these three women, he refused to pay; the pole was carved with the hope of encouraging reimbursement. The theme of debt runs through many shame poles, and some are even referred to as "debtor poles."[1] More recently, in 2007, Mike Webber carved a shame totem that was put up in Cordova, Alaska, to expose Exxon for failing to make payments for the *Valdez* oil spill. The totem includes the upside-down face of former longtime Exxon CEO Lee Raymond, sporting a Pinocchio-like nose. Webber's pole to shame Exxon made me think about the possibility of an entire pole dedicated to shaming the worst corporations.

Just as clans had specific symbols that were appropriated and turned against them, corporations have logos and iconography that have monopolized our attention and, because of their familiarity, can also be used against them. Using the logos of the companies with the greatest numbers of votes, I worked with 3-D animator Oscar Baechler to create the Shame Totem v.2.0. The result was a garish, digitally rendered 3-D shame totem, which was presented at an event at London's Serpentine Gallery in 2011. There, I had the chance to meet and collaborate with artist Brian Eno, whose sound track for the shame totem combined slowed-down indigenous hymns, financial data from NPR, and his own creations. In January 2013, I again surveyed five hundred Americans, using a different list of the biggest corporations; this time I included only U.S. corporations (so no BP) that were still in business (so no Enron). There were differences in which corporations came out as the most shameful: Walmart took an early lead, and Apple moved onto the pole. This time, Brendan O'Neill Kohl (who created all the images for this book) rendered Shame Totem v.2.1, featured here.

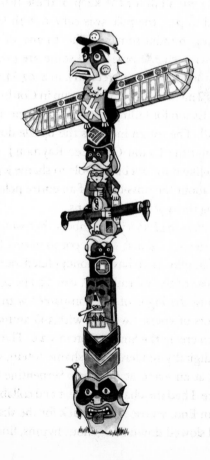

Acknowledgments

First I acknowledge you, the reader, for supporting an enterprise that requires participation and effort. A serious thank-you for your attention. I am grateful to have had some early readers who were kind, entertaining, and caustic enough to make this book better. Dalton Conley, Brendan O'Neill Kohl, and James B. MacKinnon read chapters of the book, while Paul Smaldino (thank you), Kate Barrett (thank you, thank you), and Nick Lepard (triple thank you and more) read this book in its entirety. George Dyson, one of the few people whose wisdom you'd want as much when writing as when shipwrecked, read this book, suggested many sources, and also really made the whole thing possible.

This book is in your hands rather than left as a 385KB file on my desktop thanks to the larger-than-life efforts of my friends and agents John Brockman, Katinka Matson, Max Brockman, and Russell Weinberger. It took this form due to the wonderful and talented people at Pantheon who outperformed any software. My editors Dan Frank and Jeff Alexander at Pantheon and Helen Conford at Penguin UK knew that for me to get this done required the right mixture of privacy and encouragement. I thank them all.

I wrote this book because I struggle with my own profound guilt and sadness over what humans have done to the planet and its inhabitants. This led to an interest in guilt as an environmental and then a broader phenomenon, and guilt is what led me to shame, which then led to two lines of research during graduate school: fisheries and shame. If this combination seems strange, it is because I was indulged by both institutions and people who allowed me to pursue my interests—the first being my parents, who supported all my early experiments and accepted my views even when they became different from their own. The University of British Columbia and New York University are both part of a rare ecology where well enough remains left alone. I exceeded the boundaries of fortune in getting to do my graduate research with fisheries biologist Daniel Pauly, whose world-class mind and remarkable soul guided and encouraged me through my dissertation and beyond. Christoph Hauert was a wonderful colleague at UBC who, along with Arne Traulsen and Manfred Milinski, both at the Max Planck Institute for Evolutionary Biology, were co-authors on the shame experiments. I am also thankful for the time all the people I interviewed took out of their lives to speak to me about their work.

We are all biased by our cultures and individual experiences, and I am grateful to this research for showing me this even more clearly. I hope it's obvious when the views are my own and when they are more broadly held, scientific ones, but for the record, all opinions expressed are mine. All mistakes that remain are mine, too. I am offering up an interpretation that, like shame itself, is not perfect or unchanging. But I do hope this book offers the outlaw of shaming and those who have dared to use it wisely a fair trial and a competent defense.

Notes

1 SHAME EXPLAINED

1. Carl D. Schneider, *Shame, Exposure, and Privacy* (New York: Norton, 1992).

2. Mary M. Herrald and Joe Tomaka, "Patterns of Emotion-Specific Appraisal, Coping, and Cardiovascular Reactivity During an Ongoing Emotional Episode," *Journal of Personality and Social Psychology* 83, no. 2 (2002): 434–50.

3. Christine R. Harris and Ryan S. Darby, "Shame in Physical-Patient Interactions: Patient Perspectives," *Basic and Applied Social Psychology* 31, no. 4 (2009): 325–34.

4. Stephen Bates, "Jonathan Franzen: Shame Made It Impossible for Me to Write for a Decade," *The Guardian*, October 29, 2010.

5. Virginia Morell, *Animal Wise: The Thoughts and Emotions of Our Fellow Creatures* (New York: Crown, 2013).

6. Polly Wiessner, "Norm Enforcement Among the Ju/'Hoansi Bushmen," *Human Nature* 16, no. 2 (2005): 115–45.

7. Robin Dunbar, "Co-Evolution of Neocortical Size, Group Size and Language in Humans," *Behavioral and Brain Sciences* 16, no. 4 (1993): 681–735.

8. Jennifer Jacquet, Christoph Hauert, Arne Traulsen, and Manfred Milinski: "Shame and Honour Drive Cooperation," *Biology Letters* 7 (2011): 899–901.

2 GUILT'S ASCENDANCY

1. June Price Tangney and Ronda L. Dearing, *Shame and Guilt* (New York: Guilford Press, 2002), 58.

2. Daniel M. T. Fessler, "Shame in Two Cultures: Implications for Evolutionary Approaches," *Journal of Cognition and Culture* 4, No. 2 (2004): 207–62.

3. Marvin Spevack, *A Complete and Systematic Concordance to the Works of Shakespeare* (Hildesheim, Germany: Georg Olms, 1968).

4. Roy F. Baumeister, Harry T. Reis, and Philippe Delespaul, "Subjective and Experiential Correlates of Guilt in Daily Life," *Personality and Social Psychology Bulletin* 21, no. 12 (1995): 1256–68.

5. Herant Katchadourian, *Guilt: The Bite of Conscience* (Stanford, Calif.: Stanford General Books, 2011).

6. Robert L. Trivers, "The Evolution of Reciprocal Altruism," *Quarterly Review of Biology* 46, no. 1 (1971): 35–57.

7. Toni M. Massaro, "Shame, Culture, and American Criminal Law," *Michigan Law Review* 89, no. 7 (1991): 1880–1944.

8. Martha Nussbaum, *Hiding from Humanity: Disgust, Shame, and the Law* (Princeton, N.J.: Princeton University Press, 2006).

9. James Q. Whitman, "What Is Wrong with Inflicting Shame Sanctions?" *Yale Law Journal* 107, no. 5 (1998): 1055–92.

10. Adam Duhachek, Shuoyang Zhang, and H. Shanker Krishnan, "Anticipated Group Interaction: Coping with Valence Asymmetries in Attitude Shift," *Journal of Consumer Research* 34, no. 3 (2007): 395–405.

11. Daniel Kahneman and Amos Tversky, "Choices, Values, and Frames," *American Psychologist* 39, no. 4 (1984): 341–50.

12. Tara L. Gruenewald, Margaret E. Kemeny, Najib Aziz, and John L. Fahey, "Acute Threat to the Social Self: Shame, Social Self-Esteem, and Cortisol Activity," *Psychosomatic Medicine* 66, no. 6 (2004): 915–24.

13. Michael Lewis, "Self-Conscious Emotions: Embarrassment, Pride, Shame, and Guilt," in *Handbook of Emotions,* ed. Michael Lewis and Jeannette M. Haviland-Jones (New York: Guilford, 1993), 353–64.

14. Jonathan Haidt and Dacher Keltner, "Culture and Facial Expression: Open-Ended Methods Find More Expressions and a Gradient of Recognition," *Cognition and Emotion* 13, no. 3 (1999): 225–66.

15. Peter De Jong, Madelon L. Peters, and David De Cremer, "Blushing May Signify Guilt: Revealing Effects of Blushing in Ambiguous Social Situations," *Motivation and Emotion* 27, no. 3 (2003): 225–49.

16. Charles Darwin, *The Expressions of Emotions in Man and Animals* (London: John Murray, 1872).

17. Jessica L. Tracy and David Matsumoto, "The Spontaneous Expression of Pride and Shame: Evidence for Biologically Innate Nonverbal Displays," *Proceedings of the National Academy of Sciences* 105, no. 33 (2008): 11655–60.

18. Dacher Keltner, Randall C. Young, and Brenda N. Buswell,

"Appeasement in Human Emotion, Social Practice, and Personality," *Aggressive Behavior* 23, no. 5 (1997): 359–74.

19. Dacher Keltner and Brenda N. Buswell, "Evidence for the Distinctness of Embarrassment, Shame, and Guilt: A Study of Recalled Antecedents and Facial Expressions of Emotion," *Cognition and Emotion* 10, no. 2 (1996): 155–71.

20. Michael Lewis and Douglas Ramsay, "Cortisol Response to Embarrassment and Shame," *Child Development* 73, no. 4 (2002): 1034–45.

21. Stanley T. Asah and Dale J. Biahna, "Motivational Functionalism and Urban Conservation Stewardship: Implications for Volunteer Involvement," *Conservation Letters* 5, no. 6 (2012) 470–77.

22. Alan S. Gerber, Donald P. Green, and Christopher W. Larimer, "An Experiment Testing the Relative Effectiveness of Encouraging Voter Participation by Inducing Feelings of Pride or Shame," *Political Behavior* 32, no. 3 (2010): 409–22.

3 THE LIMITS TO GUILT

1. Marion Long, "George Schaller's Grand Plan to Save the Marco Polo Sheep," *Discover*, March 2008.

2. Peter Schweizer, "Offset Away Our Guilt," *USA Today*, April 7, 2011.

3. Michael P. Vandenbergh, Thomas Dietz, and Paul C. Stern, "Time to Try Carbon Labelling," *Nature Climate Change* 1, no. 1 (2011): 4–6.

4. Stephanie Strom, "Has 'Organic' Been Oversized?" *New York Times*, July 7, 2012.

5. Nicole Charky, "Wal-Mart Removes Mislabeled Organic Products from Shelves," *Daily Finance*, April 23, 2010.

6. Keith Bradsher, "Chinese City Shuts Down 13 Wal-Marts," *New York Times*, October 10, 2011, www.nytimes.com/2011/10/11/business/global/wal-marts-in-china-city-closed-for-pork-mislabeling.html.

7. Nina Mazar and Chen-Bo Zhong, "Do Green Products Make Us Better People?" *Psychological Science* 21, no. 4 (2010): 494–98.

8. Kirk Kristofferson, Katherine White, and John Peloza, "The Nature of Slacktivism: How the Social Observability of an Initial Act of Token Support Affects Subsequent Prosocial Action," *Journal of Consumer Research* 40, no. 6 (2014): 1149–66.

9. David Sutton, "An Unsatisfactory Encounter with the MSC—A Conservation Perspective," in *Eco-Labelling in Fisheries: What Is It All About?* ed. Bruce Phillips, Trevor Ward, and Chet Chaffee (Oxford, UK: Blackwell Publishing, 2003), 114–19.

10. Shahzeen Z. Attari, Michael L. Dekay, Cliff I. Davidson, and Wändi Bruine De Bruin, "Public Perceptions of Energy Consumption and Savings," *Proceedings of the National Academy of Sciences* 107, no. 37 (2010): 16054–59.

11. Peter Whoriskey, "SUVs Lead U.S. Auto Sales Growth Despite Efforts to Improve Fuel Efficiency," *Washington Post*, December 29, 2010.

4 BAD APPLES

1. Robert Paine, "A Note on Trophic Complexity and Community Stability," *The American Naturalist* 103, no. 929 (1969): 91–93.

2. Credit for this phrase to the scientist Lewis Richardson, who published a book with this title in 1960.

3. Anthony D. Barnosky, Nicholas Matzke, Susumu Tomiya, et al., "Has the Earth's Sixth Mass Extinction Already Arrived?" *Nature* 471, no. 7336 (2011): 51–57.

4. Richard Heede, "Tracing Anthropogenic Carbon Dioxide and Methane Emissions to Fossil Fuel and Current Producers, 1854–2010," *Climate Change* 122 (2014): 229–41.

5. Naomi Oreskes and Erik M. Conway, *Merchants of Doubt: How a Handful of Scientists Obscured the Truth on Issues from Tobacco Smoke to Global Warming* (New York: Bloomsbury, 2010), 213.

6. David Berreby, *Us and Them: Understanding Your Tribal Mind* (New York: Little, Brown, 2005), 161.

7. Atul Gawande, "The Hot Spotters," *The New Yorker*, January 24, 2011.

8. Jonathan Franzen, "Emptying the Skies," *The New Yorker*, July 26, 2010.

9. Pierline Tournant, Liana Joseph, Koichi Goka, and Franck Courchamp, "The Rarity and Overexploitation Paradox: Stag Beetle Collections in Japan," *Biodiversity and Conservation* 21, no. 6 (2012): 1425–40.

10. Ian G. Warkentin, David Bickford, Navjot S. Sodhi, and Corey J. A. Bradshaw, "Eating Frogs to Extinction," *Conservation Biology* 23, no. 4 (2009): 1056–59.

11. Agnès Gault, Yves Meinard, and Franck Courchamp, "Consumers' Taste for Rarity Drives Sturgeons to Extinction," *Conservation Letters* 1, no. 5 (2008): 199–207.

12. Elena Angulo, Anne-Laure Deves, Michel Saint Jalmes, and Franck Courchamp, "Fatal Attraction: Rare Species in the Spotlight," *Proceedings of the Royal Society B* 276, no. 1660 (2009): 1331–37.

13. Bryan L. Stuart, Anders G. J. Rhodin, L. Lee Grismer, and Troy Hansel, "Scientific Description Can Imperil Species," *Science* 312, no. 5777 (2006): 1137.

14. Bernard W. Powell, "A Problem in Archaeology Too," *Science* 313, no. 5789 (2006): 916.

15. Ernst Fehr and Simon Gachter, "Altruistic Punishment in Humans," *Nature* 415 (2002): 137–40.

16. Xiao-Ping Chen and Daniel G. Bachrach, "Tolerance of Free Riding: The Effects of Defection Size, Defection Pattern and Social Ori-

entation," *Organizational Behavior and Human Decision Processes* 90, no. 1 (2003): 139–47.

17. Christel G. Rutte and Henk A. M. Wilke, "Goals, Expectations and Behavior in a Social Dilemma Situation," in *Social Dilemmas,* ed. Wim Liebrand, David Messick, and Henk Wilke (Elmsford, N.Y.: Pergamon Press, 1992), 289–305.

18. See, for instance, Will Felps, Terence R. Mitchell, and Eliza Byington, "How, When, and Why Bad Apples Spoil the Barrel: Negative Group Members and Dysfunctional Groups," *Research in Organizational Behavior* 27 (2006): 175–222. Felps also discussed his experiment with Ira Glass on the *This American Life* episode "Ruining It for the Rest of Us," which first aired December 19, 2008.

19. President George W. Bush, "Letter to Senators Hagel, Helms, et al.," press release, February 13, 2001.

20. Note, however, that a 2009 survey of 888 Americans and thousands of Europeans found overwhelming support for their respective countries doing as much as they can to fight climate change, even if other countries do less.

21. Norbert L. Kerr, Ann C. Rumble, Ernest S. Park, et al., "How Many Bad Apples Does It Take to Spoil the Whole Barrel? Social Exclusion and Toleration for Bad Apples," *Journal of Experimental Social Psychology* 45, no. 4 (2009): 603–13.

5 HOW NORMS BECOME NORMAL

1. Alan Gerber and Todd Rogers, "Descriptive Social Norms and Motivation to Vote: Everybody's Voting and So Should You," *Journal of Politics* 71, no. 1 (2009): 178–91.

2. Robert B. Cialdini, Linda J. Demaine, Brad J. Sagarin, et al., "Managing Social Norms for Persuasive Impact," *Social Influence* 1, no. 1 (2006): 3–15.

3. James J. Choi, David Laibson, Brigitte Madrian, and Andrew Metrick, "Plan Design and 401(k) Savings Outcomes," *National Tax Journal* 57, no. 2 (2004): 275–98.

4. Martin Luther King, "Nonviolence and Racial Justice," *Christian Century,* February 6, 1957.

5. Cited in Samual Bowles and Herbert Gintis, *A Cooperative Species: Human Reciprocity and Its Evolution* (Princeton, N.J.: Princeton University Press, 2011).

6. Maciek Chudek and Joseph Henrich, "Culture-Gene Coevolution, Norm-Psychology and the Emergence of Human Prosociality," *Trends in Cognitive Sciences* 15, no. 5 (2011): 218–26.

7. Elif Batuman, *The Possessed: Adventures with Russian Books and the People Who Read Them* (New York: Farrar, Straus and Giroux, 2010).

8. Nichola J. Raihani and Tom Hart, "Free-Riders Promote Free-Riding in a Real-World Setting," *Oikos* 119, no. 9 (2010): 1391–93.

9. Kees Keizer, Siegwart Lindenberg, and Linda Steg, "The Spreading of Disorder," *Science* 322, no. 5908 (2008): 1681–85.

10. Maciek Chudek and Joseph Henrich, "Culture-Gene Coevolution, Norm-Psychology and the Emergence of Human Prosociality," *Trends in Cognitive Sciences* 15, no. 5 (2011): 218–26.

11. Jennifer N. Engler and Joshua Landau, "Source Is Important When Developing a Social Norms Campaign to Combat Academic Dishonesty," *Teaching of Psychology* 38, no. 1 (2011): 46–48.

12. Peng Gong, "Cultural History Holds Back Chinese Research," *Nature* 418, no. 7382 (2012): 411.

13. Eric Posner, "Symbols, Signals, and Social Norms in Politics and the Law," *Journal of Legal Studies* 27, no. S2 (1998): 765–97.

14. Christine Ingebritsen, "Norm Entrepreneurs: Scandinavia's Role in World Politics," *Cooperation and Conflict* 37, no. 1 (2002): 11–23.

15. Robert Metz, "Market Place; Aiding Hostile Takeover Bids," *New York Times,* December 28, 1981.

16. David Skeel, "Shaming in Corporate Law," *University of Pennsylvania Law Review* 149 (2001): 1811–68.

17. Joseph Henrich, Jean Ensimger, Richard McElreath, et al., "Markets, Religion, Community Size, and the Evolution of Fairness and Punishment," *Science* 327, no. 5972 (2010): 1480–84.

18. Uri Gneezy and Aldo Rustichini, "A Fine Is a Price," *The Journal of Legal Studies* 29, no. 1 (2000): 1–17.

19. Armin Falk and Nora Szech, "Morals and Markets," *Science* 340, no. 6133 (2013): 707–11.

20. Roland Fryer, "Financial Incentives and Student Achievement: Evidence from Randomized Trials," *Quarterly Journal of Economics* 126, no. 4 (2011): 1755–98.

21. Michael J. Sandel, "What Isn't for Sale?" *The Atlantic,* February 27, 2012.

22. Josep Call, Juliane Bräuer, Juliane Kaminski, and Michael Tomasello, "Domestic Dogs (*Canis Familiaris*) Are Sensitive to the Attentional State of Humans," *Journal of Comparative Psychology* 117, no. 3 (2003): 257–63.

23. Arata Kochi, "Tuberculosis Control—Is DOTS the Health Breakthrough of the 1990s?" *World Health Forum* 18, no. 3-4 (1997): 225–32.

24. Azim F. Shariff and Ara Norenzayan, "God Is Watching You: Priming God Concepts Increases Prosocial Behavior in an Anonymous Economic Game," *Psychological Science* 18, no. 9 (2007): 803–809.

25. Jesse M. Bering, "The Folk Psychology of Souls," *Brain and Behavioral Sciences* 29, no. 5 (2006): 453–62.

26. Max Ernest-Jones, Daniel Nettle, and Melissa Bateson, "Effects of Eye Images on Everyday Cooperative Behavior: A Field Experiment," *Evolution and Human Behavior* 32 (2011): 172–98.

27. Damien Francey and Ralph Bergmüller, "Images of Eyes

Enhance Investments in a Real-Life Public Good," *PLOS ONE* 7, no. 5 (2012): e37397.

28. Kevin J. Haley and Daniel M. T. Fessler, "Nobody's Watching? Subtle Cues Can Affect Generosity in an Anonymous Economic Game," *Evolution and Human Behavior* 26, no. 3 (2005): 245–56.

29. Adriaan R. Soetevent, "Anonymity in Giving in a Natural Context—A Field Experiment in 30 Churches," *Journal of Public Economics* 89, no. 11-12 (2005): 2301–23.

30. Barry Webb and Gloria Laycock, *Reducing Crime on the London Underground: An Evaluation of Three Pilot Projects,* Crime Reduction Unit Paper 30 (London: HMSO, 1992).

31. Dominique J.-F. de Quervain, Urs Fischbacher, Valerie Treyer, et al., "The Neural Basis of Altruistic Punishment," *Science* 305, no. 5688 (2004): 1254–58.

32. Glenn Greenwald, "The Untouchables: How the Obama Administration Protected Wall Street from Prosecutions," *Guardian,* January 23, 2013.

33. Jesse Eisenger, "Why Only One Top Banker Went to Jail for the Financial Crisis," *New York Times Magazine,* April 30, 2014, http://www.nytimes.com/2014/05/04/magazine/only-one-top-banker-jail-financial-crisis.html?_r=0.

34. Hank Davis, "Theoretical Note on the Moral Development of Rats (*Rattus Norvegicus*)," *Journal of Comparative Psychology* 103, no. 1 (1989): 88–90.

35. Michael Lachmann and Carl T. Bergstrom, "The Disadvantage of Combinatorial Communication," *Proceedings of the Royal Society of London B* 271, no. 1555 (2004): 2337–43.

36. Sievert Rohwer, "Status Signaling in Harris Sparrows: Some Experiments in Deception," *Behaviour* 61, no. 1-2 (1977): 107–29.

6 THE 7 HABITS OF HIGHLY EFFECTIVE SHAMING

1. Benjamin Alamar and Stanton A. Glantz, "Effects of Increased Social Unacceptability of Cigarette Smoking on Reduction in Cigarette Consumption," *American Journal of Public Health* 96, no. 8 (2006): 1359–63.

2. Alan S. Gerber, Donald P. Green, and Christopher W. Larimer. "Social Pressure and Voter Turnout: Evidence from a Large-Scale Field Experiment," *American Political Science Review* 102, no. 1 (2008): 33–48.

3. Costas Panagopoulos, "Affect, Social Pressure and Prosocial Motivation: Field Experimental Evidence of the Mobilizing Effects of Pride, Shame and Publicizing Voting Behavior," *Political Behavior* 32 (2010): 369–86.

4. Kenneth Roth, "Defending Economic, Social and Cultural Rights: Practical Issues Faced by an International Human Rights Organization," *Human Rights Quarterly* 26, no. 1 (2004): 63–73.

5. Brian Elbel, Glen B. Taksler, and Tod Mijanovich, "Promotion of Healthy Eating Through Public Policy: A Controlled Experiment," *American Journal of Preventive Medicine* 45, no. 1 (2013): 49–55.

6. Pirjo Pietinen, Liisa M. Valsta, Tero Hirvonen, and Harri Sinkko, "Labelling the Salt Content in Foods: A Useful Tool in Reducing Sodium Intake in Finland," *Public Health Nutrition* 11, no. 4 (2007): 335–40.

7. Saira Mohamed, "Shame in the Security Council," *Washington University Law Review* 90, no. 4 (2013): 1191.

8. Stephen J. Dubner, "Another Way to Encourage Voting," *Freakonomics,* November 17, 2006, www.freakonomics.com/2006/11/17/another-way-to-encourage-voting.

9. Alexander Dyck, Natalya Volchkova, and Luigi Zingales, "The Corporate Governance Role of the Media: Evidence from Russia," *Journal of Finance* 63, no. 3 (2008): 1093–1135.

10. Polly Wiessner, "Norm Enforcement Among the Ju/'hoansi Bushmen," *Human Nature* 16, no. 2 (2005): 115–45.

11. Michiko Kakutani, "Is Jon Stewart the Most Trusted Man in America?" *The New York Times,* August 15, 2008.

12. David A. Skeel, "Shaming in Corporate Law," *University of Pennsylvania Law Review* 149 (2001): 1811–68.

13. Tiffany C. Davenport, Alan S. Gerber, Donald P. Green, et al., "The Enduring Effects of Social Pressure: Tracking Campaign Experiments over a Series of Elections," *Political Behavior* 32 (2010): 423–30.

7 THE SCARLET INTERNET

1. I am very grateful to George Dyson for his insights into the history and future of digital technology. See also *Turing's Cathedral* (New York: Pantheon 2012).

2. Marshall McLuhan, *The Gutenberg Galaxy* (Toronto: University of Toronto Press, 1962).

3. Associated Press, "Checking Cheats: China Plans Marriage Database," *The China Post,* January 6, 2011, www.chinapost.com.tw/china/national-news/2011/01/06/286518/Checking-cheats.htm.

4. Milan Kundera, *Testaments Betrayed: An Essay in Nine Parts,* trans. Linda Asher (New York: Perennial, 1996), 257–58.

5. Toni Massaro, "The Meanings of Shame: Implications for Legal Reform Psychology," *Psychology, Public Policy, and Law* 3, no. 4 (1997): 645.

6. Laura Holson, "The New Court of Shame Is Online," *New York Times,* December 26, 2010.

7. "Top 500 Delinquent Taxpayers," State of California Franchise Tax Board, www.ftb.ca.gov/aboutFTB/Delinquent_Taxpayers.shtml.

8. Thomas Moyher and Robert T. Szyba, "From the Rat to the Mouse: How Secondary Picketing Laws May Apply in the Computer Age," *Hofstra Labor and Employment Law Journal* 26 (2008): 271–300.

9. Mitchell Kapor, "Civil Liberties in Cyberspace: When Does Hacking Turn from an Exercise of Civil Liberties into Crime?" *Scientific American*, September 1991.

10. Barb Darrow, "Huffington Post to End Anonymous Comments," Gigaom, gigaom.com/2013/08/21/huffington-post-to-end -anonymous-comments/.

11. Michele Ybarra, Marie Diener-West, and Philip J. Leaf, "Examining the Overlap in Internet Harassment and School Bullying: Implications for School Intervention," *Journal of Adolescent Health* 41: S42–S50.

12. Kipling D. Williams, Christopher K. T. Cheung, and Wilma Choi, "Cyberostracism: Effects of Being Ignored over the Internet," *Journal of Personality and Social Psychology* 79, no. 5 (2000): 748–62.

13. Janis Wolak, Kimberly J. Mitchell, and David Finkelhor, "Does Online Harassment Constitute Bullying? An Exploration of Online Harassment by Known Peers and Online-Only Contacts," *Journal of Adolescent Health* 41 (2007): S51–S58.

14. Alexander Staller and Paolo Petta, "Introducing Emotions into the Computational Study of Social Norms: A First Evaluation," *Journal of Artificial Societies and Social Simulation* 4, no. 1 (2001).

15. In a UK amusement arcade, a researcher watched 303 slot machine players for four six-hour periods. Only nine slot machine gamblers were aggressive. Of the aggression, 38.2 percent was directed verbally toward the slot machine, 37.6 percent was directed physically toward the slot machine, 10.7 percent was directed verbally toward arcade staff, and 13.5 percent toward other players. Adrian Parke and Mark Griffiths, "Aggressive Behaviour in Slot Machine Gamblers: A Preliminary Observational Study," *Psychological Reports* 95, no. 1 (2004): 109–14.

8 SHAMING IN THE ATTENTION ECONOMY

1. Richard A. Lanham, *The Economics of Attention: Style and Substance in the Age of Information* (Chicago: University of Chicago Press, 2007).

2. Alex Callinicos, *Theories and Narratives: Reflections on the Philosophy of History* (Durham, N.C.: Duke University Press, 1995).

9 REACTIONS TO SHAMING

1. Frances Wilson, *How to Survive the* Titanic, *or The Sinking of J. Bruce Ismay* (London: Bloomsbury, 2011). This is a beautifully told biography of Ismay alongside a convincing literary critique suggesting that Joseph Conrad's fourth novel, *Lord Jim*, published in 1900, foreshadowed the whole woeful Ismay debacle.

2. Norbert L. Kerr, Ann C. Rumble, Ernest S. Park, et al., "How Many Bad Apples Does It Take to Spoil the Whole Barrel? Social Exclu-

sion and Toleration for Bad Apples," *Journal of Experimental Social Psychology* 45, no. 4 (2009): 603–13.

3. Redouan Bshary, "Biting Cleaner Fish Use Altruism to Deceive Image-Scoring Client Reef Fish," *Proceedings of the Royal Society London B* 269, no. 1505 (2002): 2087–93.

4. Adam Waytz and Liane Young, "The Group Member Mind Tradeoff: Attributing Minds to Groups Versus Group Members," *Psychological Science* 23, no. 1 (2012), 77–85.

5. Milton Friedman, "The Social Responsibility of Business Is to Increase Its Profits," *New York Times Magazine*, September 13, 1970.

6. Alan G. Sanfey, James K. Rilling, Jessica A. Aronson, et al., "The Neural Basis of Economic Decision-Making in the Ultimatum Game," *Science* 300, no. 5626 (2003): 1755–58. Also see James K. Rilling, Alan G. Sanfey, Jessica A. Aronson, et al., "The Neural Correlates of Theory of Mind Within Interpersonal Interactions," *NeuroImage* 22, no. 4 (2004): 1694–1703.

7. Mascha van 't Wout, René S. Kahn, Alan G. Sanfey, and André Aleman, "Affective State and Decision-Making in the Ultimatum Game," *Experimental Brain Research* 169 (2006): 564–68.

8. Olwen A. Bedford, "The Individual Experience of Guilt and Shame in Chinese Culture," *Culture & Psychology* 10, no. 1 (2004): 29–52.

9. Kalvero Oberg, "Crime and Punishment in Tlingit Society," *American Anthropologist* 36, no. 2 (1934): 145–56.

10. Michael Grabell and Sebastian Jones, "Off the Radar: Private Planes Hidden from Public View," ProPublica.

11. Jennifer L. Jacquet and Daniel Pauly, "Trade Secrets: Renaming and Mislabeling Seafood," *Marine Policy* 32 (2008): 309–18.

12. Graeme Wood, "Scrubbed," *New York Magazine*, June 16, 2013.

13. Seshadri Tirunillai and Gerard J. Tellis, "Does Chatter Really Matter? Dynamics of User-Generated Content and Stock Performance," *Marketing Science* 31, no. 2 (2012): 198–215.

14. Toni Massaro, "Shame, Culture, and American Criminal Law," *Michigan Law Review* 89, no. 7 (1991): 1880–1944.

10 THE SWEET SPOT OF SHAME

1. Aldo Leopold, "On a Monument to the Pigeon," written in 1947 and published in *A Sand County Almanac* (London: Oxford University Press, 1949).

APPENDIX: SHAME TOTEM V.2.1

1. Kalvero Oberg, "Crime and Punishment in Tlingit Society," *American Anthropologist* 36, no. 2 (1934): 145–56.

Index

Page numbers in *italics* refer to illustrations.

A NOTE ON THE TYPE

This book was set in Monotype Dante, a typeface designed by Giovanni Mardersteig (1892–1977). Conceived as a private type for the Officina Bodoni in Verona, Italy, Dante was originally cut only for hand composition by Charles Malin, the famous Parisian punch cutter, between 1946 and 1952. Its first use was in an edition of Boccaccio's *Trattatello in laude di Dante* that appeared in 1954. The Monotype Corporation's version of Dante followed in 1957. Although modeled on the Aldine type used for Pietro Cardinal Bembo's treatise *De Aetna* in 1495, Dante is a thoroughly modern interpretation of the venerable face.

Composed by North Market Street Graphics,
Lancaster, Pennsylvania

Printed and bound by RR Donnelley,
Harrisonburg, Virginia

Designed by Cassandra J. Pappas